S0-CJN-551

# MANUFACTURING IN TECHNOLOGY EDUCATION

***EDITORS***

*Dr. Richard D. Seymour*
*Associate Professor*
*Ball State University*

*Dr. Ray L. Shackelford*
*Professor*
*Ball State University*

## 42nd Yearbook, 1993

*Council on Technology Teacher Education*

**GLENCOE**

Macmillan/McGraw-Hill

Lake Forest, Illinois  Columbus, Ohio  Mission Hills, California  Peoria, Illinois

Send all inquiries to:
GLENCOE DIVISION
Macmillan/McGraw-Hill
3008 W. Willow Knolls Drive
Peoria, IL 61615

Printed in the United States of America.

***Orders and requests for information about cost and availability of yearbooks should be addressed to the company.***

---

***Request to quote portions of yearbooks should be addressed to the Secretary, Council on Technology Education, in care of the publisher, for forwarding to the current Secretary.***

*This publication is available in microform.*

University Microfilms International
300 North Zeeb Road
Dept. P.R.
Ann Arbor, MI 48106

ISBN 0-02-677140-3

# FOREWORD

Technology Teacher Education (TTE) is facing an exciting, yet challenging time in the history of the profession. One of the tasks of TTE is to redirect the curriculum to reflect contemporary Technology Education (TE).

This yearbook is the third of a series that deals with the general implementation of and instructional strategies for communication, transportation, manufacturing, and construction. Previous yearbooks *(Implementary Technology Education* and *Instructional Strategies for Technology Education)* established the setting for this series of yearbooks.

Manufacturing has and will always remain important in the evolution of technology in society. The editors have noted that: "Manufacturing involves the transformation of raw materials into industrial and consumer goods. In today's global economy, manufacturing is the basis for national wealth and power." Hence, ". . . it is critical that our citizenry understand the fundamentals, operations, and impacts of manufacturing in modern society." A contemporary study of manufacturing technology ". . . involves the organization of industrial enterprises, the processing of materials, and the development and management of industrial systems."

The first two chapters of this yearbook review the rationale and structure for studying manufacturing, and provide a societal perspective of manufacturing technology. The next four chapters provide direction for implementing manufacturing technology at the elementary, middle and high school levels, and in TTE programs. Chapters seven and eight establish the manufacturing teaching/learning environment and the facilities for teaching manufacturing. The final chapter synthesizes the yearbook. It is understood that everyone in society should understand the importance of manufacturing systems, even in an "information age."

On behalf of the CTTE members, I commend Drs. Richard Seymour and Ray Shackelford for doing an excellent job in editing this valuable yearbook. Appreciation is also expressed to the Glencoe Division of Macmillan/McGraw-Hill for their continued support of the publication of the CTTE yearbooks.

Everett N. Israel

*Everett N. Israel*
*President, CTTE*

# YEARBOOK PLANNING COMMITTEE

# *OFFICERS OF THE COUNCIL*

***President***

Everett N. Israel
Department of Industrial Technology
Eastern Michigan University
Ypsilanti, MI 48197

***Vice-President***

Anthony E. Schwaller
Department of Industrial Studies
St. Cloud State University
St. Cloud, MN 56301

***Recording Secretary***

Betty L. Rider
College of Education
The Ohio State University
Columbus, OH 43210

***Membership Secretary***

Gerald L. Jennings
Department of Business and Industrial Education
Eastern Michigan University
Ypsilanti, MI 48197

***Treasurer***

Emerson Wiens
Department of Industrial Technology
Illinois State University
Normal, IL 61761

***Past President***

R. Thomas Wright
Department of Industrial Technology
Ball State University
Muncie, IN 47306

# YEARBOOK PROPOSALS

Each year, at the ITEA International Conference, the CTTE Yearbook Committee reviews the progress of yearbooks in preparation and evaluates proposals for additional yearbooks. Any member is welcome to submit a yearbook proposal. It should be written in sufficient detail for the committee to be able to understand the proposed substance and format. Fifteen copies of the proposal should be sent to the committee chairperson by February 1 of the year in which the conference is held. Below are the criteria employed by the committee in making yearbook selections.

*CTTE Yearbook Committee*

## CTTE Yearbook Guidelines

**A. Purpose:**

The CTTE Yearbook Series is intended as a vehicle for communicating education subject matter in a structured, formal series that does not duplicate commercial textbook publishing activities.

**B. Yearbook topic selection criteria:**

An appropriate Yearbook topic should:

1. Make a direct contribution to the understanding and improvement of technology teacher education.
2. Add to the accumulated body of knowledge of the field.
3. Not duplicate publishing activities of commercial publishers or other professional groups.
4. Provide a balanced view of the theme and not promote a single individual's or institution's philosophy or practices.
5. Actively seek to upgrade and modernize professional practice in technology teacher education.
6. Lend itself to team authorship as opposed to single authorship.

Proper yearbook themes *may* also be structured to:

1. Discuss and critique points of view which have gained a degree of acceptance by the profession.
2. Raise controversial questions in an effort to obtain a national hearing.
3. Consider and evaluate a variety of seemingly conflicting trends and statements emanating from several sources.

**C. The yearbook proposal:**

1. The Yearbook Proposal should provide adequate detail for the Yearbook Planning Committee to evaluate its merits.
2. The Yearbook Proposal should include:
   a. An introduction to the topic
   b. A listing of chapter titles
   c. A brief description of the content or purpose of each chapter
   d. A tentative list of authors for the various chapters
   e. An estimate of the length of each chapter

# PREVIOUSLY PUBLISHED YEARBOOKS

*1. *Inventory Analysis of Industrial Arts Teacher Education Facilities, Personnel and Programs,* 1952.
*2. *Who's Who in Industrial Arts Teacher Education*, 1953.
*3. *Some Components of Current Leadership: Techniques of Selection and Guidance of Graduate Students; An Analysis of Textbook Emphases;* 1954, three studies.
*4. *Superior Practices in Industrial Arts Teacher Education,* 1955.
*5. *Problems and Issues in Industrial Arts Teacher Education*, 1956.
*6. *A Sourcebook of Reading in Education for Use in Industrial Arts and Industrial Arts Teacher Education,* 1957.
*7. *The Accreditation of Industrial Arts Teacher Education*, 1958.
*8. *Planning Industrial Arts Facilities,* 1959. Ralph K. Nair, ed.
*9. *Research in Industrial Arts Education*, 1960. Raymond Van Tassel, ed.
*10. *Graduate Study in Industrial Arts,* 1961. R. P. Norman and R. C. Bohn, eds.
*11. *Essentials of Preservice Preparation*, 1962. Donald G. Lux, ed.
*12. *Action and Thought in Industrial Arts Education*, 1963. E. A. T. Svendsen, ed.
*13. *Classroom Research in Industrial Arts*, 1964. Charles B. Porter, ed.
*14. *Approaches and Procedures in Industrial Arts*, 1965. G. S. Wall, ed.
*15. *Status of Research in Industrial Arts*, 1966. John D. Rowlett, ed.
*16. *Evaluation Guidelines for Contemporary Industrial Arts Programs,* 1967. Lloyd P. Nelson and William T. Sargent, eds.
*17. *A Historical Perspective of Industry,* 1968. Joseph F. Luetkemeyer Jr., ed.
*18. *Industrial Technology Education,* 1969. C. Thomas Dean and N. A. Hauer, eds. *Who's Who in Industrial Arts Teacher Education,* 1969. John M. Pollock and Charles A. Bunten, eds.
*19. *Industrial Arts for Disadvantaged Youth,* 1970. Ralph O. Gallington, ed.
*20. *Components of Teacher Education*, 1971. W. E. Ray and J. Streichler, eds.
*21. *Industrial Arts for the Early Adolescent*, 1972. Daniel J. Householder, ed.
*22. *Industrial Arts in Senior High Schools,* 1973. Rutherford E. Lockette, ed.
*23. *Industrial Arts for the Elementary School,* 1974. Robert G. Thrower and Robert D. Weber, eds.
*24. *A Guide to the Planning of Industrial Arts Facilities*, 1975. D. E. Moon, ed.
*25. *Future Alternatives for Industrial Arts,* 1976. Lee H. Smalley, ed.
*26. *Competency-Based Industrial Arts Teacher Education,* 1977. Jack C. Brueckman and Stanley E. Brooks, eds.
*27. *Industrial Arts in the Open Access Curriculum*, 1978. L. D. Anderson, ed.
*28. *Industrial Arts Education: Retrospect, Prospect*, 1979. G. Eugene Martin, ed.
*29. *Technology and Society: Interfaces with Industrial Arts*, 1980. Herbert A. Anderson and M. James Benson, eds.
*30. *An Interpretive History of Industrial Arts,* 1981. Richard Barella and Thomas Wright, eds.
*31. *The Contributions of Industrial Arts to Selected Areas of Education,* 1982. Donald Maley and Kendall N. Starkweather, eds.
*32. *The Dynamics of Creative Leadership for Industrial Arts Education,* 1983. Robert E. Wenig and John I. Mathews, eds.
*33. *Affective Learning in Industrial Arts*, 1984. Gerald L. Jennings, ed.
*34. *Perceptual and Psychomotor Learning in Industrial Arts Education*, 1985. John M. Shemick, ed.
*35. *Implementing Technology Education,* 1986. Ronald E. Jones and John R. Wright, eds.
*36. *Conducting Technical Research*, 1987. Everett N. Israel and R. Thomas Wright, eds.
*37. *Instructional Strategies for Technology Education,* 1988. William H. Kemp and Anthony E. Schwaller, eds.
38. *Technology Student Organizations*, 1989. M. Roger Betts and Arvid W. Van Dyke, eds.
39. *Communication in Technology Education,* 1990. Jane A. Liedtke, ed.
40. *Technological Literacy*, 1991. Michael J. Dyrenfurth and Michael R. Kozak, eds.
41. *Transportation in Technology Education,* 1992. John R. Wright and Stanley Komacek, eds.

---

*Out-of-print yearbooks can be obtained in microfilm and in Xerox copies. For information on price and delivery, write to Xerox University Microfilms, 300 North Zeeb Road, Ann Arbor, Michigan, 48106.

# CONTENTS

# *MANUFACTURING IN TECHNOLOGY EDUCATION*

Manufacturing involves the transformation of raw materials into industrial and consumer goods. In today's global economy manufacturing is the basis for national wealth and power. Citizens in developed nations enjoy a higher standard of living due primarily to the prosperity linked to various manufacturing industries. Even in those countries with an extensive service sector, activities such as engineering, design, retailing, inventory and accounting, and training directly support the manufacture and distribution of goods. Since many "high-tech" ventures also support manufacturing, a loss of productive industries would quickly result in a decline of personal standards. It is critical that our citizenry understand the fundamentals, operations, and impacts of manufacturing in modern society.

Currently, manufacturing is one of the major curricular organizers in technology education. The contemporary study of manufacturing technology typically involves the design of various products, the processing of materials, and the development and management of industrialized systems. This yearbook provides a blueprint for the development of quality programs in manufacturing technology education.

In the first chapter, Dr. Tom Wright reviews various structures for the study of manufacturing content. Tom has long been recognized as an international leader in manufacturing education and his perspective is unique. The structures outlined in chapter one help organize an effective study of manufacturing topics. Administrators, counselors, curriculum developers and classroom educators should be able to easily follow the sequence and scope of the manufacturing programs as described in the first chapter.

The second chapter introduces a societal perspective of manufacturing technology. The major themes reviewed in this chapter involve (a) manufacturing in a global context and (b) the impacts of the productive sector. We, as citizens, depend on various industries to provide consumer goods that fulfill our wants and needs. At the same time, our lives are greatly influenced by manufacturers in positive and negative ways. Dr. Franzie Loepp and Dr. Michael Daugherty (both of Illinois State University) have taught courses dealing with the impacts of technology, so their insights and observations in this chapter are based on their vast experiences.

The next three chapters cover the implementation of manufacturing topics at the elementary, middle school, and high school levels. The content in these sections reflect years of teaching, learning, and professional service. The authors are all successful teachers who have also contributed to numerous professional ventures (curriculum development, manufacturing in-service workshops, etc.). For instance, Mrs. Patti Farrar-Hunter has taught manufacturing courses from the 6th grade to collegiate level. Dan Chapin and Richard Otto are respected classroom educators and teacher trainers in Indiana. They understand the importance of addressing manufacturing content in public schools and recommend various strategies for teaching design, production, and enterprise topics. Dr. Ray Shackelford has helped develop secondary-level manufacturing course guides, transparency masters, and related instructional materials.

The sixth chapter addresses manufacturing education in technology teacher education programs. The future of our profession is directly linked to our ability to keep quality teachers in the public school classroom. The importance of implementing collegiate-level manufacturing experiences especially for future teachers is covered by Dr. James LaPorte of Virginia Tech. He has taught manufacturing processing, automation, and related courses at several universities and his suggestions are quite relevant. Jim also addresses the contribution of student organizational activities (such as the TECA/SME manufacturing competition) in the collegiate experience.

As educators, we often forget that our primary focus is students. Dr. Jack Wescott is an experienced methods teacher who also teaches manufacturing and construction courses. Classroom and laboratory instruction involves developing learning activities that address individual (student) needs and desires. Several of the key topics in chapter seven include evaluation methods, developing problem-solving techniques, and group dynamics.

The eighth chapter addresses the facilities and related resources required to teach manufacturing technology. Dr. Douglas Polette (Montana State University) is well known for his work in facility improvement. As a manufacturing educator and author, he has prepared many documents related to classroom and laboratory areas. This chapter outlines a model improvement plan for developing effective manufacturing programs.

The final chapter is designed to review and synthesize the content in the previous eight chapters. Although manufacturing education is not new in many schools, manufacturing programs continue to evolve in breath and scope. This section attempts to wrap-up the discussions related to addressing manufacturing in a global context. Suggestions for maintaining a dynamic program are offered with special emphasis on professional development activities.

*Richard D. Seymour and Ray Shackelford, Co-Editors*

# ACKNOWLEDGMENTS

We wish to recognize the many individuals whose contributions made this yearbook a reality. In particular, we wish to acknowledge and give credit to the chapter authors for their time, professional effort, and dedication to manufacturing programs. It is the work of committed authors that makes a task like this possible. Each author is a proven, successful classroom educator; we appreciate that each writer took valuable time to share their insights and experience with others. Tom, Franzie, Michael, Patti, Dan, Rich, Jim, Jack, and Doug—thanks!

To the members of the Council's Yearbook Committee, we extend a sincere thanks for the understanding and direction during the development of this yearbook. The committee's comments and guidance were extremely helpful. From the initial approval of the topic (in March, 1990) through final proofing of the copy, this project was a fun and rewarding experience. Again, thank you for the professional opportunity to complete this yearbook.

To the Glencoe Division of Macmillan/McGraw-Hill, we extend our sincere appreciation for the continuous commitment to technology education. The technical assistance, encouragement, and support is most important. We must thank all at Glencoe—especially Ardis, Trudy, Peggy, and Wes.

To our colleagues at Ball State University, we recognize the support and encouragement throughout this project. Our department chairman, Dr. Donald Smith, and Dean Duane Eddy, are to be commended for their assistance.

Finally, to our families and parents, we extend a large "thank you." Our professional accomplishments are a result of your support, love, and consideration. We know you realize that our service to the profession is an important part of our lives. Thank you Pam, R.C., and Nicole. Thank you Vicki.

*Richard D. Seymour*

*Ray L. Shackelford*

CHAPTER 1

# Rationale & Structures For Studying Manufacturing

Dr. R. Thomas Wright (Distinguished Professor)
*Ball State University, Muncie, IN*

The history of humankind is often presented as a series of wars, religious disputes, and governmental actions. However, in reality, it should be documented as the history of people adapting to and altering the natural environment. Osgood (1921) suggested human history started with people learning how "to talk, to worship the forces which they felt about them, to act together in groups for mutual protection, and to use their hands and minds for the better satisfying of their needs" (p. 1). Two elements in this description directly relate to developing and applying technology. First, people learned to act together and, second, they used their hands and minds to satisfy their essential needs.

The early development of technology occurred in several arenas. People learned to tame and use fire, develop stone weapons, process skins into leather, produce ceramic artifacts, weave cloth, construct dwellings, and develop means of conveyance. These and other developments dramatically changed the face of the earth. But, while many early changes were almost imperceptible, today we live in a human-dcsigned and human-built world. "Put simply, humanity now has the biblically promised dominion over the earth. We live in a managed environment. We may choose among forms of technology . . . but we may not choose to avoid technology. We have no choice but to develop our technology to greater heights of sophistication and complexity" (Walker, 1985, p. 91).

This ever-present, unavoidable, potentially beneficial phenomena called technology, is widely misunderstood, misdefined, and distrusted. To some people, *technology is hardware*. It is often associated with familiar objects such as computers, mag-lev trains, and space shuttles. To other people *technology is organization*. It is referred to as a way people arrange themselves

**Technology** is a body of knowledge and actions about....

.....applying resources
.....designing, producing, and using
.....extending the human potential
.....controlling and modifying the environment.

*Figure 1-1: A definition of technology.*

to produce useful products and services. To still other people *technology is process*. It is typically viewed as the actions used in developing, producing, and using artifacts. This third interpretation is the broadest and perhaps the most descriptive. It suggests that *technology is a body of knowledge and actions used by people to apply resources in designing, producing, and using products, structures and systems to extend the human potential for controlling and modifying the natural and human-made environment*, Figure 1-1.

The Project 2061 report (Johnson, 1989) captures the essence of technology by suggesting that it is "the application of knowledge, tools, and skills to solve practical problems and extend human capabilities" (p. 1). The report continues by stating that technology "is conceived by inventors and planners, raised to fruition by the work of entrepreneurs, and implemented and used by society."

Since everyone is affected by technology in various forms, it is essential this topic be studied by young and old alike. Our citizenry should learn how humans adapt the natural environment and fashion products and services that improve life. This yearbook will specifically focus on technologies related to the manufacturing sector. This first chapter will look at how technological topics and the content of manufacturing are often introduced in today's schools.

## STUDYING TECHNOLOGY

As defined, technology is a system and its major actions involve the acts of designing, producing, and using. The challenge for educators is to develop a curriculum that allows students to view technology realistically and consistently. A number of approaches for structuring content have been advocated by technology educators. Among the more common include the applied science approach of physical, bio-related, and information technology (Savage, 1989); the resources approach of information, energy, and matter (Wolters, 1989); and the human-productive activity approach of communication, construction, manufacturing, and transportation (Warner, 1965; Hales and Snyder, n.d.).

The human-productive approach seems to show the most promise. It suggests that people have, are, and will be actively using technology in communicating information and ideas, constructing structures, manufacturing products, and transporting people and goods, Figure 1-2.

Studying technology as a system requires that the components of the system be identified. The Jackson's Mill document (Hales and Snyder, n.d.) identified a system as a combination of elements or parts that work in an orderly, predictable way to accomplish a desired goal. The document presented the "universal systems model" as having inputs, process, output, and feedback. Each element of the system was described as follows:

Inputs: All resources needed to accomplish the goals of the system including people, knowledge, materials, energy, capital, and finance.

Process: A scheme of purposeful actions or practices that make up the technical means of the system.

Outputs: The goal or ends to which all inputs and processes (technical means) are applied.

Feedback: The mechanisms that provide preferred direction for the system.

This approach provides a way to view the dynamic phenomena called technology. However the systems model provides only part of the view. The contexts in which the system operates and is studied must also introduce the social-cultural nature of the topic. All technology impacts and is impacted by historical, economic, social, cultural, and environmental contexts. Only

Humans are productivity-involved in **applying technology** to:

**Communicate** information and ideas
**Construct** structures
**Manufacture** goods
**Transport** people and cargo

through designing, producing, and using devices and systems.

*Figure 1-2: Technology as a human-productive activity.*

when technology is viewed in relationship with the ethos in which it operates, can a realistic understanding be developed.

## Manufacturing As A Technology

Humans have used a variety of technologies throughout history. Typical examples include construction, communication, manufacturing, and transportation technologies. The productive activity of manufacturing is fundamental to all developed or developing countries. Manufacturing accounts for 80-85% of the wealth of the United States and manufacturing output was projected to have grown 15% during the 1980s alone (Weiss, 1987).

Furthermore, manufacturing around the globe is changing. Restructuring is the password to the future. This action has caused many people to lose sight of the fundamental principle of manufacturing. Put simply, manufacturing involves those actions completed in a factory to change the form of materials to add to their worth. In short, manufacturing is in the *form utility* business. People place higher commercial value on wood fibers in the form of lumber than they do trees. Likewise they value wood in the form of furniture more highly than in the form of lumber. It is the role of manufacturing to successively change raw materials into industrial materials and, then, into industrial and consumer products. Using these basic themes to describe this productive activity, manufacturing can be defined as *those actions that occur in a factory to change material resources into industrial materials and on into usable products.*

Schools should address manufacturing in this basic form. Instructional activities and entire classes ought to be implemented that cover the importance of manufacturing to a free enterprise system. Further coursework should be structured to review the acts of designing, producing, and using manufacturing technologies. These classes must also provide students with a way to learn about the manufacturing sector in a laboratory-based environment. Themes for individual activities and courses will be explained in later chapters.

# THE SCOPE OF MANUFACTURING CONTENT

Manufacturing can be narrowly conceived as processing materials into products or more broadly viewed as all actions used to change raw materials into products. This second perspective provides the most realistic view of manufacturing as a method of providing form utility. The broader view involves the technological actions of designing, producing, and using and the three major manufacturing phases of securing resources, producing industrial materials, and producing industrial and consumer products.

In addition, like all technology, manufacturing is a purposeful, human action. Therefore, it is managed. To many people management and industry are corollary terms but this is not necessarily true. It is important to describe manufacturing as both technology *and* industry. Technology is any use of technical means to extend the human potential, while industry is the societal institution that develops and produces most technology.

From a societal context, manufacturing can be personal or commercial. Manufacturing technology can be applied by an individual in avocational pursuits, by groups in a service setting (e.g., a fraternal or religious group producing items for needy children), or by an economically driven group (e.g., a Fortune 500 firm). Each of these applications will use the same two basic types of technology. They will use *process technology* (technical means) in designing and producing products and *managerial technology* to insure that the processing actions are efficient and appropriate.

Most likely, these technologies will differ only in sophistication as they are applied in the personal or industrial settings. The sketches used by the hobbyists may be replaced by CAD-generated drawings in a commercial enterprise. Likewise the hand-crafted product built by an artisan is comparable to the machine-produced, affordable, artifact of industry. Similarly the informal cost controls used by individuals will be replaced with corporation cost accounting, purchasing departments, and glossy annual reports.

The interface between the technological actions, manufacturing phases, and the societal context provides the fundamental structure for an adequate study of manufacturing. This relationship is depicted in the array shown in Figure 1-3. Students of manufacturing should be exposed to this broad view as they study the organization, activities, and impacts of the productive sector.

## Technological Actions

All technological artifacts are the result of human volition. They were designed because someone thought people either (a) needed the devices or system or (b) a desire to possess them could be developed. The responsible development and use of technology moves through the three major actions including technological actions, manufacturing phases, and the societal context.

Applying this three-phase approach to the study of manufacturing involves using three separate systems—the *design* system or approach, the production system (*produce*), and the consumption or application system (*use*). This approach can be illustrated in a model in which resources flow

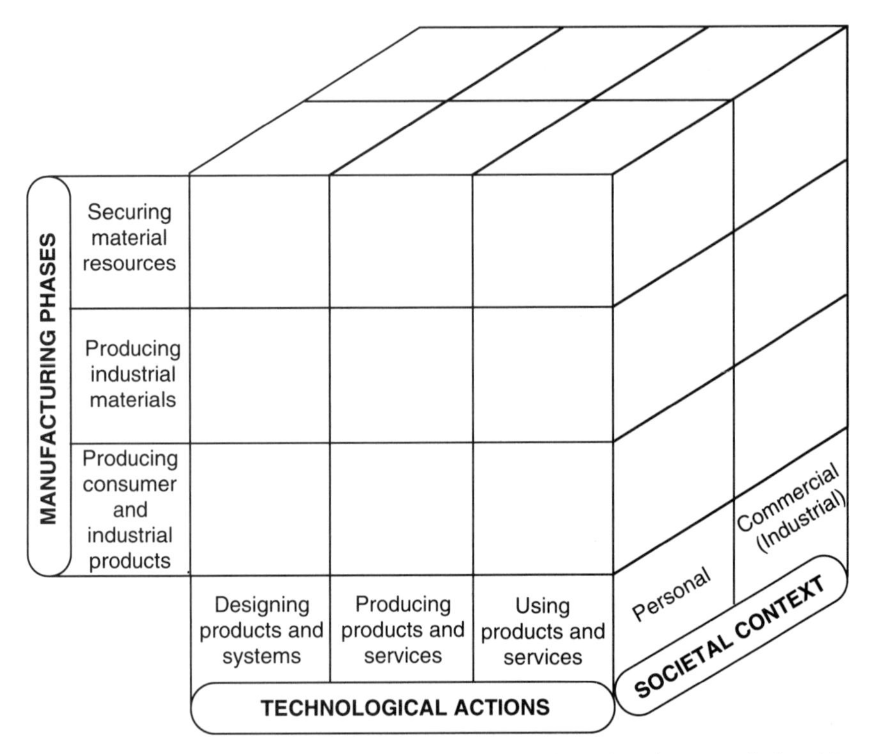


*Figure 1-3: The scope of manufacturing technology involves a relationship between its phases, technological actions, and societal contexts.*

into the manufacturing system. The model shows how products are designed, produced, and made available to people. These products are used by people to meet individual or group needs. At each step of the design-produce-use cycle various impacts are felt both within the system and in the larger personal, societal, and environmental arena, Figure 1-4.

Design actions govern the way the product is produced and likewise how the production facility will impact how the product is to be made. Also, the customer and the product's intended use will directly influence both designing and producing the product. Also, each design, produce, and use action will directly impact the environment, resource consumption, quality of life, and numerous related factors.

Each action has its own unique technological system that applies resources to reach a goal. These fundamental actions are the essence of technology. They are the application of technical means to extend the human potential.

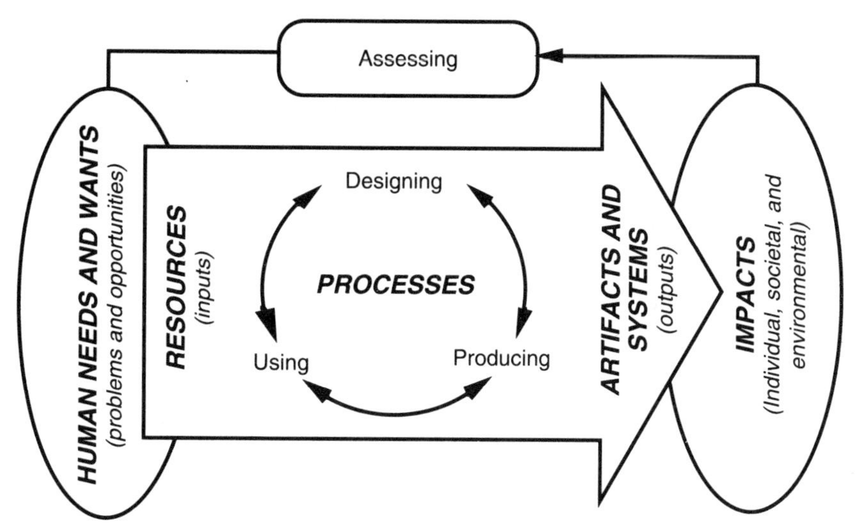


*Figure 1-4: Manufacturing, like all technological systems, includes the actions of designing, producing, and using resources.*

*Designing Products and Systems.* Technological devices, as well as systems, are designed to meet human needs and wants. Therefore, one phase of manufacturing deals with *the procedure used to address technological problems and opportunities*. This procedure describes how new and improved products and manufacturing systems are created. It is perhaps best compared to the scientific method in that discipline.

Over time this procedure has been described as the design method (Lindbeck, 1963), problem-solving method (Waetjen, 1989), and the technological method. A common outline for this process includes:

1. Defining the problem.
2. Developing alternate solutions.
3. Selecting a solution.
4. Implementing and evaluating the solution.
5. Redesigning the solution.
6. Interpreting the solution (Savage and Sterry, 1990).

The design process includes both divergent and convergent thinking processes. First, designers apply divergent thinking to generate multiple solutions to the problem or opportunity. Then, they switch to convergent thinking as they narrow the selection to what appears to be the best suggestion and develop this optimal solution. This process is uniquely different from the work of the creative artist. The artist tries to convey meaning through artistic media with little regard to cost and market restrictions. In contrast, the designer typically works under economic, technical, and market restrictions. This procedure, as shown in Figure 1-5, describes the way the human-made world is created through discovery, invention, innovation, and development.

*Producing Products.* The actions that make manufacturing a unique endeavor are the processes used to produce products. These involve the three stages of manufacturing already introduced: securing materials resources, producing industrial materials, and producing consumer and industrial goods.

In modern manufacturing, raw materials are converted into useful products following a carefully engineered sequence of activities. Component parts are produced, then combined into sub-assemblies and final (completed) products. Organizational skills and a knowledge of material processing (described later in this chapter) are essential to a successful venture.

*Using Manufactured Products.* Manufacturing does not stop with the transformation of raw materials into products. It extends into properly using manufacturing products. Each industrial and consumer good completes a five-step cycle of use (often referred to as a product's "life cycle"), Figure 1-6. The consumer must first select the appropriate product among those that will meet the basic need or requirement. Then they operate the product

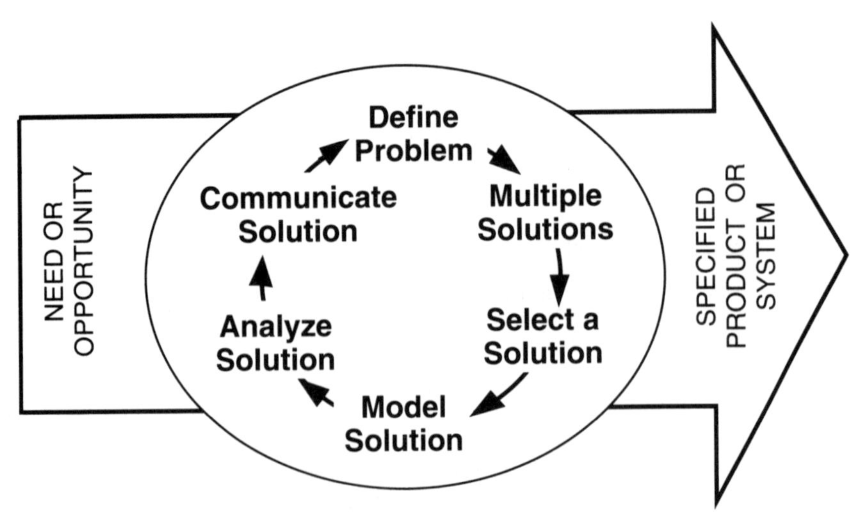


*Figure 1-5: Designing technological products and systems.*

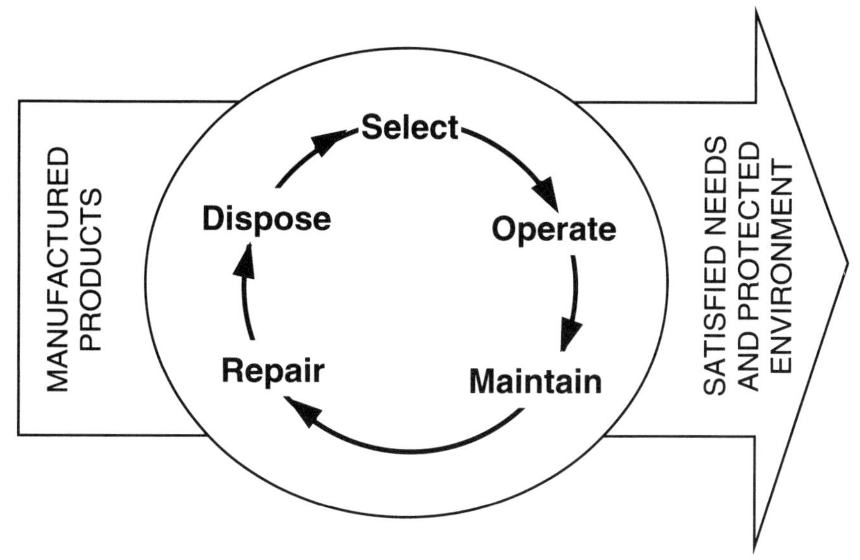


*Figure 1-6: The 5-step cycle for using manufactured products.*

and provide it with routine service. Poorly functioning products can either be maintained, repaired, or discarded. Responsible consumers then properly dispose of products by recycling the materials, if possible, or placing the product in a suitable disposal site.

## Manufacturing Phases

An understanding of manufacturing processes is built upon two basic bodies of knowledge. The first is the knowledge of materials. The physical sciences describe materials as having three basic states: gas, liquid, and solid. All three of these are used in manufacturing. The second area involves the manipulation of materials (i.e., processing) using the principles of chemistry, physics, or biology. Products that have structure and form are manufactured from solid or engineering materials. For instance, chemical products are produced in refineries using a distillation (thermal) process. In a similar way, conditioning and assembly processes also change the nature of the materials involved, Figure 1-7.

## Materials

The knowledge of materials includes information about types and properties, Figure 1-8 (Wright, 1990, 1992, 1993). All materials can be grouped by type into one of three basic categories:

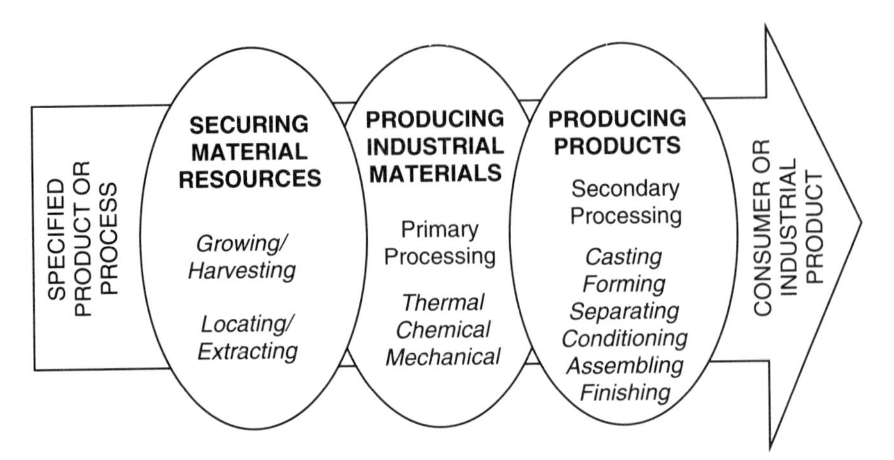


*Figure 1-7: The phases of manufacturing technology.*

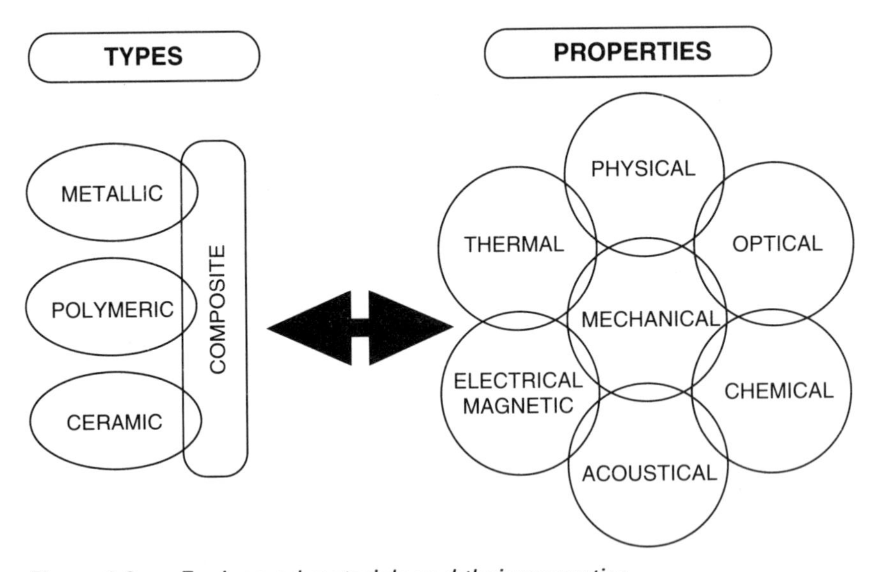


*Figure 1-8: Engineered materials and their properties.*

Metallic: Inorganic, crystalline substances made up of unit cells which are generally divided into ferrous and nonferrous categories.

Polymeric: Organic, non-crystalline hydrocarbon substances including the thermosets, thermoplastics, and elastomers.

Ceramic: Mostly inorganic, crystalline materials including the clays, cements, refractories, and abrasives.

Additionally, some materials are categorized as composites which include all combination materials that are composed of a matrix and filler. These materials usually have the most desirable of several different qualities.

Materials are selected for use based upon their properties. The vast number of properties are often grouped into one of the following seven properties:

| | |
|---|---|
| Mechanical properties: | Describes the ability to support mechanical forces or loads. |
| Physical properties: | Describes the appearance and texture. |
| Thermal properties: | Describes the reaction to changes in temperature. |
| Chemical properties: | Describes the reaction to chemicals. |
| Electrical/magnetic properties: | Describes the reaction to electrical and magnetic forces. |
| Optical properties: | Describes the reaction to light waves. |
| Acoustical properties: | Describes the reaction to sound waves. |

*Securing Material Resources.* All manufactured products have their roots in a material resource. Typically, these inputs are classified as either exhaustible or renewable resources. Exhaustible resources have a finite quantity available and cannot be replaced by human or normal natural action. Typical exhaustible resources include fossil fuels, metallic ores, clays, salts, and other natural minerals. Renewable resources are derived from living plant and animal organisms and include wood, vegetable fibers (cotton, flax, etc.), animal hides and hair (leather, wood, etc.), and grains. Renewable resources are genetic materials and are sometimes called bio-materials.

Obtaining exhaustible materials resources involves an entirely different procedure than obtaining renewable resources, Figure 1-9. Exhaustible resources must first be located then extracted from the earth. Mining or drilling techniques are often used to extract these resources. In contrast, renewable resources are first developed through natural or human-managed planting/breeding activities. The plants or animals are then grown to maturity, followed by harvesting, slaughtering, or similar activities.

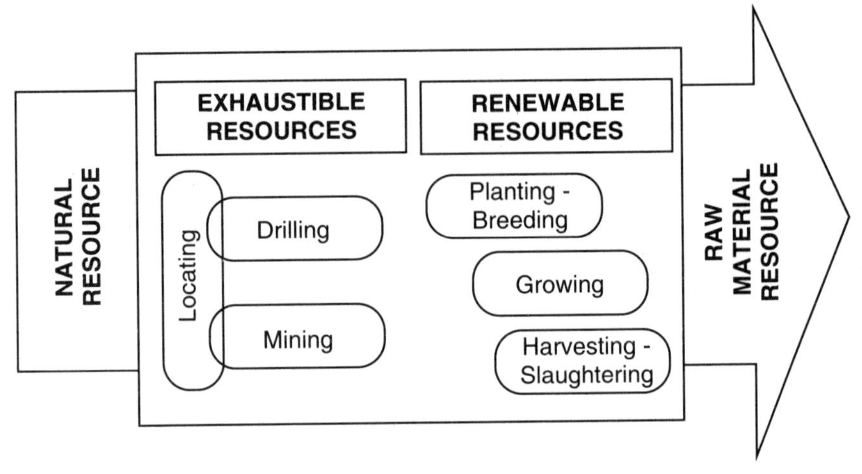


*Figure 1-9: Processes involved in securing and developing material resources.*

*Producing Industrial Materials.* In a manufacturing system, material resources are the key manipulated input. The first step usually involves changing raw materials into a usable form (referred to as industrial materials or standard stock). This transformation takes place in primary processing facilities such as steel mills, lumber mills, and petroleum refineries.

The primary processes, as seen in Figure 1-10, used to convert the materials can be grouped under three major headings:

Thermal processes: Using heat to alter the structure or chemical composition of materials such as when smelting or baking.

Chemical processes: Using chemical reactions to alter the structure and composition of a material such as the processes found in a petrochemical plant. A sub-set is electrochemical processing which uses electricity to enhance the chemical action such as in aluminum refining.

Mechanical processes: Using mechanical force to alter a raw material by crushing, sawing, and slicing.

*Producing Industrial and Consumer Products.* At some point materials are transformed into a final form that serves an industrial and consumer end use. In this form, the material is called a product and is made up of component parts and/or sub-assemblies.

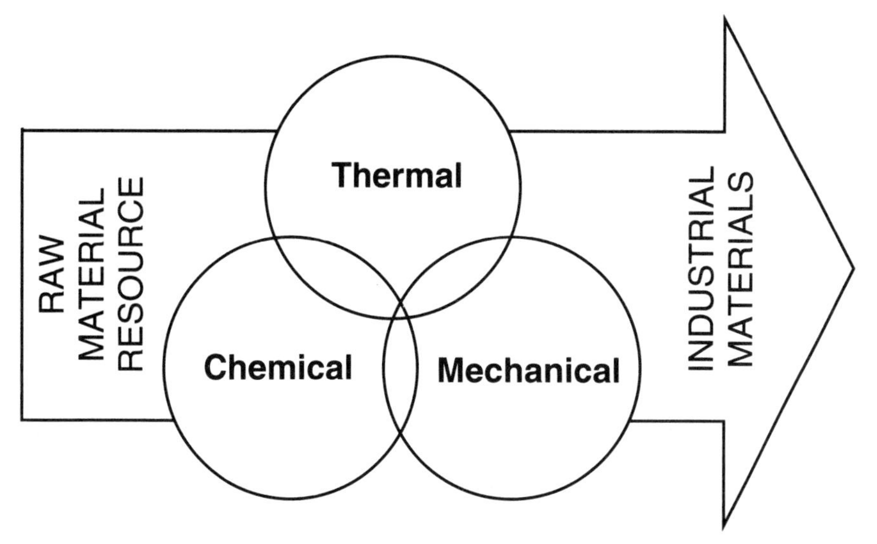


*Figure 1-10: Primary processes used to convert raw materials into industrial materials.*

The production of products is the result of secondary processing activities. These activities provide the final size, shape, and surface finish to the material(s). Secondary processes are numerous and varied; however, as shown in Figure 1-11, they can be grouped into six basic processes:

1. Casting & Molding: The manufacturing process whereby the desired material is liquified, then introduced into a previously prepared mold cavity of proper design. The material is allowed to solidify in the mold before being extracted.

2. Forming: The manufacturing process in which the size and shape, but not the material volume, of a part is changed by the application of a force. This force is above the yield strength (i.e., the point at which the material will not return to its original state) and below the fracture strength.

3. Separating: The manufacturing process in which excess material is removed to change the size, shape or surface finish of a part.

4. Conditioning: The manufacturing process in which the properties of a material are changed by application of heat, chemicals, or stress.

5. Assembly: The manufacturing process in which two or more parts are temporarily or permanently fastened together.

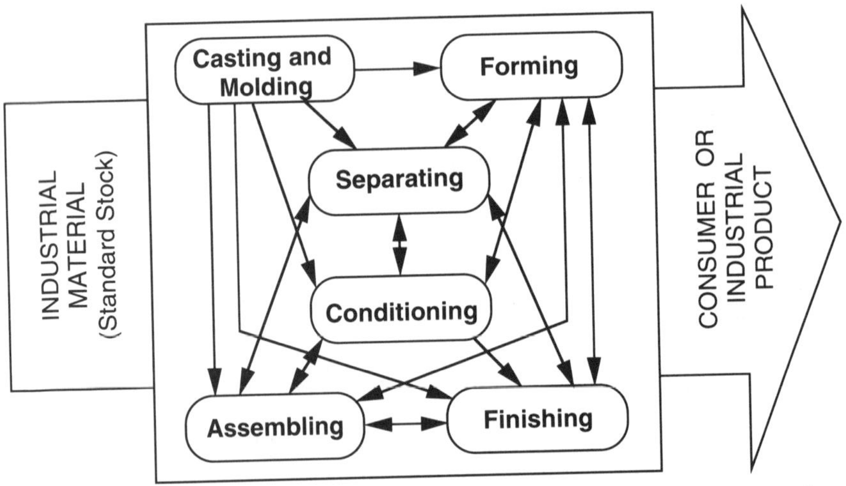


*Figure 1-11: Secondary processes used to convert industrial materials into products.*

6. Finishing: The manufacturing process by which a material's surface is beautified and/or protected. (Wright, 1990, 1993)

## Management

Manufacturing is more than materials and processes. It should be viewed as a managed system that converts materials into useful products to meet human needs and wants. The system is designed to capture a creative idea and transform it systematically into a tangible item. To accomplish this task requires management. Fundamentally, management involves all actions that insure that the manufacturing activities are efficient and appropriate. The difference between managing at the individual versus corporate level is a matter of societal context.

Management of manufacturing enterprises involves performing a set of *functions* with a given *level of authority* and responsibility in a specific *area of activity*. The function describes what management does on a routine basis. The four typical functions of management are:

Planning: Setting goals and a course of action.

Organizing: Structuring the job(s) into manageable tasks.

Directing: Assigning tasks and supervising their completion.

Controlling: Comparing results against the outlined plan.

Generally, managers have an identified level of authority and responsibilities assigned to them. These managers operate within an area of activity within an enterprise. The areas illustrated in Figure 1-12 include:

| | |
|---|---|
| Research and Development: | Discovers, develops, and specifies characteristics for new products and processes. |
| Production: | Engineers the manufacturing facility and produces scheduled products to stated quality standards. |
| Marketing: | Identifies the market for the product and promotes, sells, and distributes the product. |
| Industrial Relations: | Develops programs to recruit, select, and train needed workers. Promotes positive relations between the company and its employees, public, and union. |
| Financial Affairs: | Raises and controls the company's money and purchases the necessary equipment, materials, and supplies. |

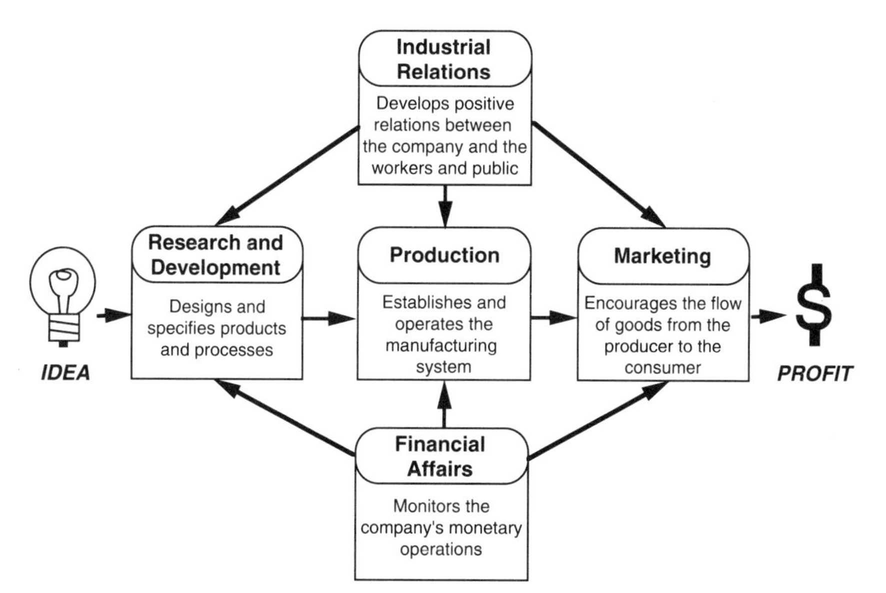


*Figure 1-12: The managed areas of activity.*

# SUMMARY

Manufacturing includes those activities that add value to materials by changing their form and creating products. Manufacturing technology, regardless of its application, uses material processing technology and management technology to transform material resources into products that people need and/or want. Material processing, in its broadest sense, includes all activities used to (a) obtain material resources, (b) produce industrial materials, and (c) create industrial and consumer products. These activities may be practiced on a personal or commercial level. Most successful products are produced by larger industrial corporations.

Manufacturing starts when natural (exhaustible) resources are located or genetic (renewable) resources start their developmental cycle. It continues as these resources are obtained through mining, drilling, or harvesting actions. In-plant conversion of the material resources commences with primary processing. During this phase, material resources are converted into industrial materials or standard stock. These materials are the inputs to secondary processing activities that ultimately result in industrial and consumer products.

In addition, manufacturing systems are purposeful, managed activities. They rely on research and development activities to create products and processes; production activities to generate products; and marketing activities to promote, sell, and distribute the outputs of the system. This product evolution process is supported by financial affairs and industrial (employee) relation activities.

An adequate study of manufacturing should involve a review of the technological actions, manufacturing phases, and managerial context. At each step in the design-produce-use cycle in manufacturing technology, environmental, societal, and personal impacts are considered. The ultimate goal is to help students understand how manufacturing technologies are applied to create products that make life more comfortable for people while protecting the natural environment and societal structure.

# *REFERENCES*


Hales, J., & Snyder, J. (n.d.). *Jackson's mill industrial arts curriculum theory*. Charleston: West Virginia Department of Education.

Johnson, J. R. (1989). *Technology*. Washington, D.C.: American Association for the Advancement of Science.

Lindbeck, J. (1963). *Design textbook*. Bloomington, IL: McKnight and McKnight.

Osgood, E. L. (1921). *A history of industry*. Boston: Ginn and Company.

Savage. E. (1989). *A model for technology education in Ohio*. Bowling Green, OH: The Model Technology Systems Project.

Savage, E., & Sterry, L. (1990). *A conceptual framework for technology education*. Reston, VA: International Technology Education Association.

*Science for all Americans*. (1989). Washington, D.C.: American Association for the Advancement of Science.

Waetjen, W. (1989). *Technological problem-solving*. Reston, VA: International Technological Education Association.

Walker, D. W. (1985). *Curriculum and technology: Current thought on curriculum*. Alexandria, VA: Association for Supervision and Curriculum Development.

Warner, W. E. (1965). *A curriculum to reflect technology*. Columbus, OH: Epsilon Pi Tau.

Weiss, J. (1987). *The future of manufacturing*. Red Bank, NJ: Bus Fac Publishing.

Wolters, F. (1989). A PATT study among 10- to 12-year-olds. *Journal of Technology Education*, *1*(1), 22-33.

Wright, R. T. (1987). *Processes of manufacturing*. South Holland, IL: Goodheart-Willcox.

Wright, R. T. (1990). *Manufacturing systems*. South Holland, IL: Goodheart-Willcox.

Wright, R. T. (1992). *Technology systems.* South Holland, IL: Goodheart-Willcox.

Wright, R. T. (1993). *Exploring manufacturing.* South Holland, IL: Goodheart-Willcox.

CHAPTER 2

# Manufacturing Technology: A Societal Perspective

Dr. Franzie Loepp (Distinguished Professor)

and

Dr. Michael Daugherty (Assistant Professor)
*Illinois State University, Normal, IL*

> A strong manufacturing sector is absolutely vital to progressive nations. There is an irrefutable correlation between the growth of a country's manufacturing industry and its long-term standard of living. It is manufacturing that drives the economies of such nations. . . the wealth creating activities are those that produce something, like mining, farming, and in particular, manufacturing. . . You cannot have an ongoing service economy without a manufacturing base (Barcus, 1992, p. 40).

Manufacturing is inexplicably linked with other segments of society, such as transportation, communication, education, and public service. Thus, manufacturing has a major global, national, regional and local influence on the way all citizens live. Because of the vital linkages to society, there is enormous competition among municipalities, states, and nations for the introduction of new manufacturing enterprises into their economy.

The economic viability of a community is often directly related to the success of local manufacturing industries. Manufacturing firms provide direct employment in the transportation of materials and products, production sector, and the marketing of raw materials, pre-processed materials, and saleable products. Furthermore, those directly engaged in manufacturing activities also require services which in turn provide additional employment within a community.

Manufacturing also has a regional impact. A major manufacturing enterprise can attract a large number of smaller supporting companies into a region. This is particularly true during an era when just-in-time (JIT) manufacturing has become so prevalent. Manufacturers using JIT principles often allow no more than a thirty-minute inventory for some components and sub-assemblies. Suppliers find it necessary to be located near their primary customers. This mushrooming effect can have an enormous impact on a regional economy. Therefore, a world-class competitive manufacturing corporation can have a far-reaching influence on the economic viability of a state, province, or region.

Manufacturing firms also have a tremendous impact at the national level. Whether it be the gross domestic product (GDP), national security, balance of trade, or the evaluation/devaluation of its currency, manufacturing plays an important role. Historically, this has been demonstrated by the economic growth of various countries. Faltermayer (1990) notes that "beginning in the late nineteenth century and throughout most of the twentieth century, the United States was a leading industrial power, making it possible for the standard of living to nearly double with each generation" (p. 44). During the last half of the twentieth century, however, a number of other nations including Germany, Korea, Japan, etc. have become increasingly competitive in the manufacturing sector. As their manufacturing base grew, the standard of living of their citizens also increased while the standard of living in the U.S. leveled off. This trend illustrates the importance of manufacturing to a nation's standard of living as well as the country's emergence as a global power.

To capitalize on the advantages afforded by international cooperation, various free trade agreements are being negotiated. One example of this is the formation of the European Economic Community (ECC)—an organization of several nations in Western Europe. Cooperative ventures such as this are destined to have a strong impact on manufacturing industries and the worldwide economy. Governments and multinational corporations will continue to seek innovative ways to join economic forces and expand their production sectors.

## MANUFACTURING IN TODAY'S GLOBAL SOCIETY

Since manufacturing plays such an important role within the global, national, regional and local economies, governments and multinational corporations are continuously faced with difficult decisions. On the one hand, there are many incentives for expanding the manufacturing base, while on the other hand, there are hazards to such growth. For instance, everyone realizes the importance of protecting people, the earth, and its natural resources (Poling, 1991). As a result, governments have responded by developing regulatory agencies such as the U.S. offices of Occupational Safety and Health Administration (OSHA) and the Environmental Protection Agency (EPA). International associations have also formed which strive to protect the environment. One example is Green Peace, an international association devoted to protecting the environment and natural wildlife. Further, the United Nations has focused on international issues that merit global attention (e.g., the condition of the ozone layer).

All of these endeavors have a significant impact on the worldwide status of the manufacturing sector. This section will review how governments strive to regulate manufacturing firms while promoting fair competition and national interests. The business practices of multinational corporations will also be discussed (as they apply to international competitiveness).

## Government Policies and Regulations

Due to the far-reaching effect the manufacturing industry has on society, it is natural for government to establish policies which promote manufacturing growth. Tax incentives, trade agreements, and manufacturing loans are some common governmental incentives. However, in capitalistic societies where profit is a primary motive, it is also important to regulate manufacturing in order to protect the environment as well as those employed in manufacturing industries. Governments, as well as international businesses, are beginning to take a longer view of the impacts of manufacturing, and they are therefore, trying to overcome the technical and cost-related challenges of being kinder and gentler to the planet (Poling).

Although government regulations are often viewed as having a detrimental effect on business/industry, they are generally written to have long-term positive impacts. For example, the U.S. Occupational Safety and Health Administration (OSHA) regulations are designed to assure a safe work environment. Over the longterm, the cost of providing a safe environment is lower than the cost of dealing with human injury, suffering, and downtime. A prime example is when an automobile manufacturer, scheduled to produce one unit per minute, has to stop production due to an accident. The lost productivity may exceed $10,000 per minute. They also face the probability of increased accident insurance premiums and perhaps the hiring and training of a new worker. Collectively, these costs can be prohibitive and the expenditure does not add value to the product. So, the extra money expended to engineer the plant to reduce the risk of accidents is money well spent.

Likewise, regulations regarding the discharge of hazardous waste (with oversight provided by the federal EPA) can also produce long-term savings. It is much easier to control the immediate discharge of hazardous waste than to take care of the problem once it has been discharged into the environment. Governmental policies and agencies attempt to insure that manufacturers are responsible for their own actions.

Sometimes it is politically expedient for government to protect certain segments of the manufacturing industry by establishing tariffs and quotas. U.S. aircraft industries are demanding legislation to protect their global interests due to the success of the Airbus family of aircraft. Airbus, composed of companies in France, Germany, the United Kingdom, and Spain have been increasing their market share partly due to huge government subsidies and tax incentives that protect the European industries. British Aerospace, for example, received $725 million from the British government to design and develop a new wing for their aircraft. Yet, U.S. aircraft manufacturers receive no direct funding or tax relief (March, 1990) from the federal government. Although there is a trend for some politicians to support protectionism legislation, others lean toward and support free trade (Holzinger). Ultimately, however, trade barriers in the form of quotas and tariffs cost the consumer by allowing lower quality and higher cost products to saturate the market. Additionally, they tend to allow industries to become less efficient. Beams and Motes (1991) suggest that the trade barriers "cost American consumers ten times more than they help American business. Such practices cost the average U.S. family $1,200 a year in needlessly high prices" (p. 30).

Governments protect long-term manufacturing investments in many ways. One familiar technique is by maintaining an office that issues patents and trademarks. The protection of innovative ideas provides an incentive for corporate research and development (R & D), which leads to improved materials, processes, and products. It is important to honor the concept of a patent on a worldwide basis.

Trademarks are also beneficial to the manufacturing industry. They can serve to establish a global identity for companies and thereby, increase the potential market for their products. These, too, should be honored on a worldwide basis because counterfeit products can have an adverse effect on the market potential of the original product. Also, counterfeit products seldom measure up in terms of quality and dependability, thus providing the consumer with an inferior product.

## Modern Business Practices

Manufacturing industries around the world are constantly forced to re-evaluate their business practices. For instance, many manufacturers are changing their environmental policies due to public and private pressures. Conoco Incorporated has begun buying only double-hulled tankers to safeguard against oil spills. The Atlantic Richfield Company (ARCO) claims it "has unveiled a cleaner replacement for leaded gasoline. By selling 840,000 gallons of the new blend a day, ARCO claims to have cut air pollution in the Los Angeles basin by 120 tons per day" (Ivey, Grover, Therrien, Shao, April 23, 1990, p. 99). While environmental regulations (such as taxes related to the volume of pollution produced, emission standards, and recycling requirements) are sometimes viewed as being detrimental to economic growth, in the long run they are necessary (Holzinger, 1991). Additionally, many companies have become more pro-active in their attitude towards pollution control and environmental clean-up. Oil companies in the U.S. spent more than $3 billion in hazardous waste cleanup and pollution charges in 1989; few industries can absorb this type of expense in today's competitive marketplace.

Cultural issues and business practices can either promote or hinder growth in manufacturing industries. Manufacturers who wish to expand their markets to other countries *must* learn their language and study their culture. A product name in one country may have a negative connotation in another language. For example, a General Motors automobile named "Nova" did not sell very well in Spanish-speaking countries because the term means "it does not go" in Spanish. Perhaps a low-cost modification, such as a different nameplate, is needed to enhance a product or service.

Traditionally, U.S. executives have expected foreign counterparts to negotiate and provide contracts in English. Unless bilingual experts are involved in the negotiation process, clear communication between the parties involved is often compromised resulting in misunderstanding. A better situation is for all parties to study the business practices, cultural traits, and experiences of the partners involved.

Another example of a societal condition that impedes international trade is the reluctance of the U.S. society to totally adopt the International System of Units referred to as the *Metric System of Measurement*. Since all other major industrialized countries have adopted this system, it is natural for them to expect the system to be used on a universal basis. It also impedes the sale of U.S. products abroad. The situation relates closely to the phenomenon of "nationalism" or the strong tendency for a people of a nation to put the well-being of their own country ahead of the world. Extreme patriotism can blind a country's view of the advantages that can be gained from international cooperation.

As the twenty-first century approaches, two seemingly opposing forces seem to be converging. On one hand, the democratization of governments in the formerly socialistic eastern block countries has given rise to extreme patriotism on the part of many ethnic groups and factions. Simultaneously, western European countries are anticipating an exponential growth in economic power through the formation of the European Economic Community. While some details are yet to be determined, the goal of the EEC is to eliminate long existing trade barriers and allow for the free flow of people, goods, capital, and services, thereby, creating the world's largest single free market economy (Poling). Other examples of international cooperation include the development of the Canada/United States Free Trade Agreement (FTA) to promote manufacturing in North America (Gold & Leyton-Brown, 1988). The objective of the FTA is to secure access of Canadian markets and investments for U.S. manufacturers, and vice versa. The United States is also negotiating a similar free trade agreement with its neighbor to the south, Mexico. U.S. domestic labor-intensive industries oppose FTA agreements because without them U.S. manufacturers have access to low-cost labor in neighboring countries. Other industries support FTA because they tend to increase market potential and product lines.

Currently, a vast majority of international trade and investment is controlled by multinational firms based in the ECC, the United States, and/or Japan (Gold & Leyton-Brown). Multinational companies strive for access to the markets provided by these major industrial powers because people in these countries have the resources and a standard of living which allows them to purchase a wide range of products and services.

Another similar activity is the multinational research initiative involving government and private sector cooperation from the United States, Europe, and Japan. This initiative is referred to as the Intelligent Manufacturing Systems (IMS) project, which is destined to break new ground for international development and cooperation. Heaton (1991) notes how the "IMS project is striving to harmonize world-wide standards and promote the diffusion of new technologies" (p. 36). But the dream of a global economy may still be years away since Naisbitt & Aburdene (1990) remind us that "for a global economy to work, however, there must eventually be completely free trade among all nations" (p. 4).

Free trade among all nations does present some rather unique problems. One problem with international interdependence in the manufacturing sector involves the national security issues. For example, since the military is a heavy consumer of modern technology, there is the fear among developed nations that the tendency for increased nationalism may warrant the withholding of important technologies from the militaries of economic partners. Alarmists fear that countries with manufacturers who are in the forefront of a given technology (e.g., advanced microprocessors) will hold an inordinate amount of power and leverage in an international conflict. Others have a much more positive view. They emphasize that interdependence can be a strong force for causing governments to solve differences through negotiation rather than violence, since the destruction of an economic partner would have an adverse effect on their own economy.

# CONSEQUENCES OF MANUFACTURING

Since the beginning of the Industrial Revolution, every country that has delved into the manufacturing enterprise has seen a marked increase in their standard of living. With the increased globalization and technological advancement of the manufacturing sector, it is easy for an outside viewer to glorify the benefits of manufacturing and overlook the consequences. However, looking at only the benefits of technology and not its consequences is a mistake society can scarcely afford to make. For instance, one may think the computer has only positive things to offer manufacturing and society. However, the computer has introduced a series of negative side effects, not the least of which is Carpal Tunnel Syndrome which is crippling millions of office workers. In order to further examine the consequences of manufacturing, it is imperative that one examine the planned and unplanned ramifications of those endeavors, as well as the immediate and delayed responses to manufacturing technology, Figure 2-1.

## Planned and Unplanned Impacts of Manufacturing

"When the space shuttle, Challenger, burst into flames less than two minutes into its lift-off, people around the world were forced to grasp the intensity of an unexpected tragedy. Stunned in silence, it is likely that many lost all faith in technology—even if just momentarily" (Markert, 1989, p. 130). Tragedies such as the space shuttle disaster force onlookers to examine the planned and unplanned consequences of modern technology. It is highly unlikely that the disaster would have occurred had designers, engineers, and the launch teams all understood the influence of the cold weather prior to the launch. It has been proven that the extreme cold was partially responsible for the O-ring failure which ultimately caused the explosion in the vehicle.

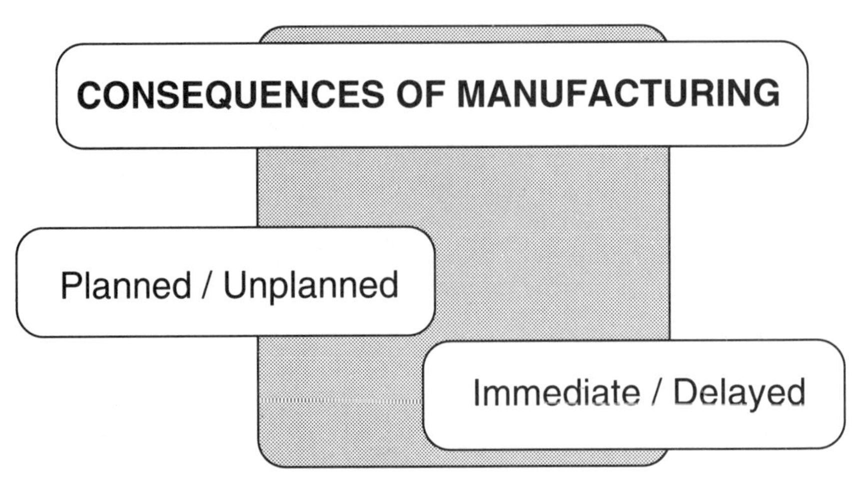


*Figure 2-1: Two ways to classify the consequences of modern manufacturing.*

Conversely, many products designed by NASA have also resulted in positive, unplanned consumer products. The Apollo lunar suit worn by a dozen moon-walking Astronauts was a masterpiece of design and engineering. The technology used in these suits and other space-related products has spawned a series of new consumer products. One little-known feature of the suit was the use of special three-dimensional "spacer" materials in the lunar boots for cushioning and ventilation. That material has turned up, in modified form, as the key element of a new family of athletic shoes designed for improved shock absorption, energy return, and reduced foot fatigue (NASA, p. 56).

In their haste to develop new and more advanced products, manufacturers often overlook potential, unplanned hazards. A simple example of this is the common lawn mower. The lawn mower has evolved from a manually pushed, mechanical device which seldom harmed anyone into a power-driven device that requires increasingly less effort. Unfortunately, Markert (1989) notes that "currently, 50,000 U.S. citizens are injured each year while working with their lawn mowers" (p. 195). Therefore, the consequences of the lawn mower injuries probably were not taken into full consideration.

Planning is under way at the Center for Robotics at the University of Cincinnati to develop a robotic lawn mower which may solve some of the inherent safety problems with lawn mowers. Researchers at the Center believe that the first domestic robot may be a lawn mower. The robotic lawn mower prototype which currently is in existence includes a warning light bumper that stops the machine, sonar detectors that locate moving objects, omni-directional vision systems that enable the mower to avoid obstacles and monitor progress, and a complete computer guidance system (Tyre, 1991).

Long-term planning seems to be the most appropriate route to follow in order to avoid the negative unplanned consequences of manufacturing. However, no amount of planning can avoid all negative consequences. In December, 1982, a factory in Alton, Pennsylvania, installed a new high-precision machining line that promised to move the plant into the modern age of metalworking. Managers hoped to eventually convert all of the plant's production lines to a similar technology. After four years of planning, development, and efforts to "debug" the system with little success, no one was sure whether quality had improved appreciably. Despairingly, company managers decided to ship the production line off to a sister plant overseas.

Some of the most striking examples of planned and unplanned consequences of manufacturing are occurring in developing countries. In their haste to enter the global market, many developing countries overlook the unplanned consequences of manufacturing. The People's Republic of China is presently embarking on an aggressive journey in the direction of modernization. The Republic is allowing greater initiative among the localities, thus allowing for the expansion of local resources (energy and raw materials) and manufacturing talent (entrepreneurial skills). However, all of these initiatives appear to be on a collision course with the traditional Communist system of government (Markert, p. 245). Many traditional Communists are alarmed at this modernization effort and are demanding action to stop the initiatives.

India is a developing nation heavily dependent on agricultural performance and the importation of small manufacturing plants. It boasts nearly 140 centers of higher learning which graduate the third largest number of scientists and engineers in the world. The technical prowess of these scientists and engineers is evidenced by India's recent nuclear and space accomplishments. Yet, as with many developing countries, India regularly overlooks many safety and environmental concerns associated with manufacturing. This became alarmingly clear to the rest of the world when a large chemical manufacturing plant in Bhopal, India released toxic chemicals into the atmosphere killing many local residents.

Another unplanned consequence involves the source of materials, energy, and capital to support the productive sector. Manufacturing can cause a country to become economically dependent on the raw materials from other countries. Japan, a high technology manufacturing nation, imports four-fifths of its energy. "Although it has reduced oii imports by 26% since the early 1970s, it is still heavily dependent on the turbulent Middle-Eastern waters" (Markert, p. 264).

Finally, when new products reach the marketplace, their positive nature is usually accentuated and they are often hailed as technological wonders. The potentially negative effects caused by the device are often overlooked, often for many years. Perhaps the best example was the original U.S. cotton gin. Developed by Eli Whitney, the cotton gin was hailed as the savior of the cotton industry. It markedly increased the amount of cotton that could be produced by speeding up the fiber and seed separation process. But it was not realized until years later that this labor-saving device had actually increased the demand for slaves among the cotton-producing plantations. The plantations could now produce more cotton to be ginned and, therefore, needed more slaves to produce the cotton.

## The Immediate and Delayed Consequences of Manufacturing

Overlooking the long-term consequences of technology for the immediate gratification of a need can sometimes have dire consequences. Shulman (1991) reminds us of a prime example:

> "Don't plan to vacation on Johnson Island. Granted, you can't beat the remote location, 800 miles from the nearest neighbors in Hawaii, U.S.A. And, the 600 acre island is a wildlife refuge, replete with palm trees, coral formations, and white sandy beaches. However, Johnson Island may be as uninviting as an island can get. It is here that the United States is starting to incinerate the first of its huge, lethal, and aging arsenal of chemical weapons manufactured for the U.S. military over the past decades". (p. 18)

Recently, manufacturers have attempted to enhance their understanding of the immediate versus delayed consequences of manufacturing systems and products. Currently, many companies are conducting technological assessments prior to marketing new products. A technology assessment not only identifies the immediate consequences of the product, but also serves to evaluate the long-term or delayed consequences. By evaluating the delayed consequences of a product, multinational companies and manufacturers can avoid long-term environmental or health hazards associated with the items.

The trend to avoid long-term problems has already started in many industries. For instance, manufacturers strive to catch the eye of the consumers by developing flashy packaging and advertising paraphernalia which ultimately end up in landfills. As landfills become more and more congested and regions become unwilling to accept the garbage of other regions, manufacturers around the globe have begun to realize the necessity of reducing the volume of unnecessary packaging and waste material. The age of disposable manufacturing goods may be coming to a close. "Recently, the Kodak Corporation announced that it will recycle its disposable cameras, as well as the chemicals used to produce film and develop pictures" (Ivey, et al., p. 98). Similar programs at "at least 37 giant corporations including Monsanto, DuPont, Westinghouse, Allied Signal, Proctor and Gamble, Texaco, IBM, and AT&T have launched environmental initiatives" (Ivey, et al., p. 99) to promote a cleaner planet. On a positive note, these strategies can more than pay for themselves. The Union Carbide Company recycled, reclaimed, or sold 82 million pounds of waste in the first half of 1989, efforts that generated $3.5 million in income and avoided disposal costs of $8.5 million" (Schiller, 1990, p. 101).

"During the mid-1980s, the U.S. Environmental Protection Agency (EPA) estimated dangerous chemicals were leaking out of at least 16 thousand landfills throughout the U.S. countryside" (Markert, p. 315). These chemicals were produced to serve an immediate need and the delayed consequences were not addressed. As technology in manufacturing continues to escalate, it will become imperative that the immediate need of a product will have to be weighed against the delayed consequences of that product. Markert also reminds us that "a century ago, if a maintenance worker failed to attach a carriage wheel properly, the lives of a half dozen people might be threatened. Today, when a manufacturer fails to install an O-ring properly in an L-1011 aircraft, more than 300 lives are at stake" (p. 385). Thus, it is easy to see that the ramifications of manufacturing extend beyond environmental issues into both personal and corporate concerns.

As public pressure has forced manufacturers to become more environmentally conscious, they have also become more aware of the effect of their production operations on individuals, the environment, and world resources. It becomes imperative that manufacturers look to eliminate many of the negative consequences of modern processing techniques. "In recent years, acid rain has received widespread recognition as a serious environmental problem in many areas of the world, including Canada, Scandinavia, Japan, and the Northeast United States" (Markert, p. 305). Much of the acid rain is produced by the emissions of manufacturing plants. In a recent public opinion survey, three-fourths of Americans believed business and industry should be more environmentally aware, but only 36% of those surveyed think industry is doing a good job with their environmental programs.

With all of this environmental consciousness, it is recognized that some of the more serious air pollution problems created by manufacturing are not even occurring near the earth's surface (Markert, p. 307). Industrial emissions are contributing to what has been labeled the "greenhouse effect" which is creating an alarming hole in the ozone layer of the earth's atmosphere. Governmental and public pressure for environmental consciousness are causing many manufacturers to reevaluate their global production efforts. However, this "greening" of the manufacturing climate is not as evident in developing countries where government regulation and public pressure are not as great.

In the context of environmental change, multinational corporations, like all of us, are part of the problem. Corporations and individuals are also central to the solution (Ivey, et al.). "The Union Carbide Corporation from 1990 to 1994 will spend $310 million a year on the environment, including a new plastics recycling venture which recycles or reuses 50% of the hazardous waste produced by the company" (Ivey, et al., p. 97). Individuals must also become responsible for their actions when they use and dispose of manufactured items.

The positive and negative long-term consequences of manufacturing can also be viewed from an economic point of view. After an onslaught of foreign competition in past decades, many U.S. companies have devised strategies to fight competition. U.S. shoe manufacturers have approached Congress with demands of tariffs on foreign products. Many U.S. domestic electronics companies have moved operations off-shore to take advantage of the competitors low wages and benefits. Subsequently, many U.S. firms have stopped making T.V.'s and VCR's; rather, they have become distributors for products made abroad (Markert). Locating manufacturing operations abroad has many economic consequences, the least of which may be a decreased domestic tax base.

International competition, though initially seen as a negative consequence, may also provide positive long-term benefits for manufacturers. "In efforts to become more competitive, many companies have developed manufacturing systems which challenge traditional assumptions and contemporary production procedures" (Markert, p. 196). One example is the reaction of officials at Chrysler, Ford, and General Motors to global competition. The challenge from off-shore automakers prompted the Big 3 automakers to develop and implement new facilties. One area of the U.S., formerly known as the "rust belt," is now being applauded as America's "automation alley" (Markert). The term "automation alley" refers to the increased use of automated technologies and the renewed competitive nature of the manufacturers in the region.

*Figure 2-2: Several of the important trends in manufacturing technology.*

# GLOBAL TRENDS IN MANUFACTURING

"As we move toward the twenty-first century, a booming global economy will be one of the overarching trends influencing our lives" (Naisbitt & Aburdene, p. xix). Factors that contribute to this economy are often referred to as world-class manufacturing, lean production, flexible manufacturing, and time-based competition (Van, 1991). These terms refer to more than just productivity. They also refer to the trends for increased product quality, more variety, customizing, more convenience, and timeliness (Carnevale, 1991). Related trends include increased automation, a focus on education, and domination by multinational corporations, Figure 2-2.

This global economy presents governments and multinational companies with tremendous opportunities for growth and expansion into international markets. This section will review a few of the significant trends related to manufacturing technology in our global society.

## Multinational Companies

Government policy has a definite impact on the development of companies that trade in the world market. In the United States for example, the government is somewhat reluctant to become deeply involved in the private sector for fear of being accused of providing an unfair advantage for certain manufacturers. In countries such as Japan, the government cooperates fully with private industry for the purpose of helping their manufacturers compete more favorably in the global marketplace. Regardless of a governmental policy, manufacturers worldwide are realizing that it is imperative to adapt to fast-changing worldwide conditions and to develop innovative, quality products faster than their competitors (Poling).

Manufacturers keep abreast of worldwide products by forming branches in other countries or forming partnerships with similar manufacturers in foreign lands. By diversifying in this manner, corporations benefit by having local support including a knowledge of local laws, business customs, and the language (Holzinger). Additionally, these partnerships can be most helpful in assembling the local distributors for products. While these practices offer advantages to the manufacturer, they also elevate the expectations of consumers. Consumers are becoming more demanding, requiring more new products, more quickly, better quality, more value, and total satisfaction with both the sale and subsequent service (Poling).

Multinational companies also work internationally for financial advantage. By moving manufacturing processes to the location that offers the cheapest labor and/or natural resources, costs are decreased. Additionally, increased market potential offers the opportunity for the maximization of capital investment due to the principle of economics of scale (i.e., the concept that increasing the quantity of items produced will lower the cost of each individual product).

## Automation

As manufacturing industries strive to produce larger quantities of products they usually turn to automation for the answer. Automation in manufacturing helps to bring down costs and assures a consistent quality product. Familiar attempts to automate various phases of a production system include computer-aided design (CAD), computer numerical control (CNC), and flexible manufacturing systems (FMS).

Numerical control (NC) is the operation of a machine by a series of coded instructions comprised of numbers, letters of the alphabet, and other symbols. These instructions are transmitted into pulses of electrical current or other output signals that actuate motors and other devices to control an operation (Gettelman, Nordquist and Herrin, 1990). Through user-friendly software such as APT (Automatic Programming Tool), NC programmers can direct a machine to perform a specific task with a high degree of accuracy and repeatability. Relying on a computer to direct the machinery results in "computer numerical control" (or CNC).

While NC is controlled through hard-wired control units, the physical components for CNC are soft-wired. It is not the control unit elements, but rather the executive program that makes the control unit perform in a special way. The executive program is loaded into the CNC's memory by the control builder. CNC machines usually have an internal memory for program storage. It is possible to "read in" and store the information from a part program. An advantage of CNC is its ability to interface with peripheral machine features such as automatic tool changing, pallet interchange, on-line diagnostics, and automatic coolant control. Late generation CNC systems often incorporate microprocessor-based PLC (programmable logic controller) to gather information from various sensors and control auxiliary actions (Gettelman, et al.). CNC and the use of PLC's are essential to computer-integrated manufacturing (CIM).

CIM is the concept of using digital computers as the integrating force throughout the entire manufacturing process. It has become the umbrella term to include the concepts of computer-aided design/computer-aided manufacturing (CAD/CAM) as well as incorporating front-office functions such as cost accounting, payroll, and inventory control. The ultimate goal of CIM is to make all manufacturing functions interdependent and interactive. Advanced hardware and software in CIM systems can lead to the development of flexible manufacturing systems (FMS).

A flexible manufacturing system has several distinguishing characteristics including:

1) An automatic materials handling subsystem linking machines in the system and providing for automatic interchange of workpieces in each machine.

2) Automatic continuous cycling of individual machines.

3) Complete control of the manufacturing system by the host computer.

4) Is lightly manned or may have unmanned capability.

Due to its very nature, this system is capable of adapting to various models of a single product on the same manufacturing line (Gettelman, et al.).

## Quality

Quality is primary among the new competitive standards. When the United States lost its market share in computer chips, the productivity rate was the highest in the world, but the quality was suspect. U.S. chips were not as reliable as those produced in Japan (Clausing, 1989). Therefore, it was no accident that when the United States established its first major award for excellence, it was an award for *quality* rather than productivity (Carnevale). Through total quality management (TQM) principles, workers are empowered to build quality into products. Modern workers learn to use statistical quality control (SQC) techniques to determine whether or not the process being used is producing an acceptable component. Workers are also encouraged to solve problems together and to offer suggestions for continuous improvement. In addition to internal quality measures, clients are viewed as an important resource in determining product quality. Client response is taken seriously, and suggestions are integrated into the production, sales, and/or service functions.

## Variety

Variety is another trend that is having a tremendous impact on manufacturing. "Between 1979 and 1989, the number of items carried on U.S. supermarket shelves rose from 12,000 to 24,000" (Carnevale, p. 29). This trend for greater product variety is an outgrowth of increased productivity and quality. Greater efficiency has improved the standard of living for many people, so now they can afford more variety. With more attention given to the wishes of the consumer, manufacturers are utilizing flexible technologies to produce more variety with little increase in cost. Examples of flexible manufacturing technologies include computer numerical control (CNC), computer-integrated manufacturing (CIM), and flexible manufacturing systems (FMS).

## Customization

A fifth trend, customization, is closely related to variety. Through customization, the manufacturer is able to more precisely respond to the needs of consumers. Examples in this area include customized manufactured homes as ordered by the client. With the help of a computer-aided design (CAD) system, a client can design the home of their dreams. As soon as the design has been approved, digitized information is transmitted directly to the factory where modern systems cut and assemble components that are delivered to the work site as soon as excavation and foundation work is complete. Another example is in the apparel industry. One manufacturer in Japan is capable of producing customized suits within a matter of hours. A client is scanned by a laser with the information fed into a computer that utilizes this input data to design a garment that allows for an exact fit. Then, this information is fed into CNC machines that cut and assemble the product (Carnevale). The result is a customized suit in less than two hours.

## Convenience

As the pace of life increases, time becomes more and more important. Convenience can give a manufacturer a strong competitive edge over rivals in virtually any industry. Companies that have the foresight to offer significant conveniences to the consumer are likely to gain an advantage. In fact, many people are willing to pay a premium for convenience.

Quite often, convenience is designed into a product. An example might be the newer control units on microwave ovens that can be programmed so dinner is ready when arriving home from work. Or, perhaps it is a unique remote control for the operation of a television or VCR. Another form of convenience has to do with delivery. In the past few years, a number of convenience stores have opened. Although prices are higher, people are willing to pay a premium in order to avoid an additional stop. In the automotive industry, manufacturers seek to make the maintenance process as convenient as possible. Monitors provide the driver with a reminder so when service is needed, the customer is alerted (Carnevale).

## Timeliness

Timeliness is another major element that provides manufacturers with a competitive edge. Carnevale notes that "products that come to market on budget but six months late will earn 35% less profit over five years than products that come out on time but are 50% over budget" (p. 32). It is, therefore, increasingly important to have a short time from the development of a concept for a product to final delivery in the marketplace. Computer-aided design (CAD), rapid prototyping, and computer-integrated manufacturing are all techniques that are used to bring new products to the consumer more quickly. The utilization of these technologies offers a remarkable edge to manufacturers.

## Education

Finally, there is a definite trend toward the requirement for a more educated workforce within the manufacturing industry. Government has a big responsibility in the preparation of the workforce. Japan, Korea, and Germany have proven that a country with a superior labor force can be competitive on a worldwide basis, even though they lack raw materials and low cost energy. The way to keep the United States industry from being overtaken in the rising tide of world competition is to substantially improve the education of the workforce (Faltermayer). No nation has produced a highly qualified technical workforce without first providing its workers with a strong basic education. Most international general knowledge tests show U.S. children rank behind children in other industrialized countries (National Center on Education and the Economy, 1990).

Over the past decade, an increasing number of nations with excellent educational systems have emerged to challenge U.S. dominance in the manufacturing sector (Poling). Unfortunately, managers in other nations view education as an investment while, in the U.S., it is often considered as a cost (National Center on Education and the Economy). Support for public and private education is critical to national interests.

Further, U.S. industries must not wait for public schools to prepare the workforce for the twenty-first century. They must provide an education to those currently employed. The bulk of the current American workforce went to school before computers and calculators were prevalent. These workers must often be trained for even the most remedial of tasks.

# SUMMARY

Manufacturing is linked with almost every segment of modern society. A strong manufacturing base is absolutely vital to maintaining global, national, regional, and local economies. Manufacturing often creates wealth among and between nations, and likewise, the lack of manufacturing often causes intense economic difficulty. The power of manufacturing to alter an economy is so great that many associations and governments devote their entire existence to promoting and controlling it.

Modern manufacturing, however, is not perfect. Manufacturing industries are often blamed for environmental and natural resource exploitation. For this reason, many governments have developed regulatory offices such as the U.S. office of Occupational Safety and Health Administration (OSHA) and the Environmental Protection Agency (EPA). These agencies not only have the responsibility to regulate industry, but also serve to protect the environment, society, and production sector itself.

Furthermore, manufacturing is not without its consequences. While striving to solve problems the industry often creates new ones. Conversely, products manufactured for a specific purpose often yield unexpected benefits, such as products that were designed for space exploration which later were developed into consumer products. Tragedies such as the Space Shuttle disaster or the chemical accident in Bhopal, India force manufacturers and society alike to examine the planned and unplanned consequences and nature of manufacturing technology. Long-term planning seems to be the most appropriate route to follow in order to avoid the unplanned consequences and accentuate the positive consequences of manufacturing. Long-term planning can bridge the gap between evaluating the immediate and delayed consequences of manufacturing. By initially assessing a technology, manufacturers have learned they can avoid most unforeseen consequences which may prove to be economically and environmentally devastating.

Because manufacturing has such an enormous impact on society, most governments have established policies which promote growth within the sector. These policies often include tax incentives, trade agreements, and manufacturing loans. Governmental policies and regulations are sometimes viewed as protectionist acts by global competitors. By establishing tariffs and quotas for national products, governments often exclude global markets for products and may hinder the growth of specific manufacturing industries.

Governmental policy has a definite impact on the development of companies that trade on the world market. In countries such as Japan, the government cooperates fully with private industry for the purpose of helping their manufacturers compete more favorably on the global marketplace. Regardless of governmental policy, manufacturers worldwide are realizing that it is imperative to adapt to a fast-changing world and develop innovative, quality products faster than their competitors.

The global economy is one of the powerful trends affecting manufacturing today. Global manufacturing has been made possible by increased quality, variety, customization, timeliness, and the increased convenience of product lines. These new trends in global manufacturing have been heavily influenced by the integration of the automation technologies into manufacturing. Some manufacturing systems that rely heavily on the computer include computer numerical control, computer-integrated manufacturing, computer-aided drafting, computer-aided manufacturing, and flexible manufacturing systems.

Perhaps the most important element in modern manufacturing is education. Success in a technological venture involves educated management and labor. Industries spend billions on training and re-training programs. At the same time, public and private schools have been challenged to prepare young learners for a tumultuous career in a technological world.

Manufacturing, through its very existence, drives and stifles economies, develops products that protect and threaten the environment, solves problems while creating new ones, and is inextricably integrated and linked to almost every segment and fabric of our society.

# REFERENCES


Barcus, J. F. (1992). Manufacturing awareness. *Modern Applications News*, 26(4), 40-41.

Beams and motes. (1991, December 7). *The Economist*, *321*(7736), 30.

Carnevale, A. P. (1991). *American and the new economy*. Monograph of the Employment and Training Administration (Grant No. 99-6-0705-079-02). Washington, DC: U.S. Department of Labor.

Clausing, D. (1989). The semiconductor, computer and copier industries. *The Working Papers of the MIT Commission on Industrial Productivity*, (Volume 2). Cambridge: The MIT Press.

The Commission on the Skills of the American Workforce. (1990). *America's choice: High skills or low wages!* Rochester: National Center on Education and the Economy.

Faltermayer, E. (1990, September 24). Is 'made in U.S.A.' fading away? *Fortune*, 62-73.

Gold, M., & Leyton-Brown, D. (1988). *Trade-offs on free trade*. Toronto: Carswell.

Holzinger, A. G. (1991, December). Selling in the new Europe. *Nation's Business*. *79*(12). 18-24.

Ivey, M., Grover, R., Therrien, L., & Shao, M. (1990, April 23). The Greening of Corporate America. *Business Week*, 96-103.

Machine tools and their control. (1990). In K. Gettelman, W. Nordquist, & G. Herrin (Eds.), *Modern Machine Shop 1990 NC/CIM Guidebook* (pp. 51-70). Cincinnati: Modern Machine Shop.

March, A. (1990, January). The future of the U.S. aircraft industry. *Technology Review*, 27-36.

Markert, L. R. (1989). *Contemporary technology: Innovations, issues, and perspectives*. South Holland, Illinois: The Goodheart-Willcox Company, Inc.

Morrison, D. (1992, June). There and back. *The Discovery Channel Magazine*, *8*(3), 11.

Naisbitt, J., & Aburdene, P. (1990). *Megatrends 2000*. New York: Avon.

NASA. (1989) *Spinoff*. Washington: U.S. Government Printing Office.

Networking for DNC and FMS. (1990). In K. Gettelman, W. Nordquist & G. Herrin (Eds.), *Modern Machine Shop 1990 NC/CIM Guidebook* (pp. 235-250). Cincinnati: Modern Machine Shop.

Noyelle, T. (1988). *Services and the new economy: Toward a new labor market segmentation.* New York: National Center on Education and Employment, Teachers College, Columbia University.

Numerical control: What it's all about. (1990). In K. Gettelman, W. Nordquist & G. Herrin (Eds.), *Modern Machine Shop 1990 NC/CIM Guidebook* (pp. 35-46). Cincinnati: Modern Machine Shop.

Poling, H. A. (1991, November). American industry: Challenges for the 1990s. *USA Today*, *120*(2558), 22-24.

Schiller, Z. (1990, April 23). P & G tries hauling itself out of America's trash heap. *Business Week*, 101.

Shulman, S. (1991, October 18). Artificial germ warfare. *Technology Review*, 94(7).

Tyre, M. J. (1991, October). Managing innovation on the factory floor. *Technology Review*, *94*(7), 58-65.

Van, J. (1991, November 3). Old manufacturing ideas crippling U.S. *Chicago Tribune*, p. 12.

Van, J. (1991, November 4). Processes, not products, open the path to profits. *Chicago Tribune*, Section 1.

Van, J. (1991, November 5). Firms tool up with information. *Chicago Tribune*, Section 1.

CHAPTER 3

# Manufacturing Technology At The Elementary Level

Mrs. Patricia Farrar-Hunter (Technology Educator)
*Roswell, GA*

As educators, we must introduce students to manufacturing, how it relates to their lives, and the significance of exploring such a subject at an early age. This introduction should also emphasize the importance of modern manufacturing and its contribution to our technological society. But the students, young and impressionable learners, need to be taught complex topics in a simplified manner. One critical task is to define manufacturing in simplistic terms so that the elementary students can comprehend the topic.

Depending on the age of the elementary student, manufacturing can be defined in one of the following ways:

- Producing a product
- The making of products for people to use
- Changing the form of materials to add to their value, thus producing useful products

For instance, cutting a piece of wood into the shape of a car and adding two axles and four wheels greatly adds to the value of the material. It has been transformed into a useful product for a child. It has just become a manufactured product, Figure 3-1.

We all use manufactured products every day of our lives. We read books that are manufactured. We listen to cassette tapes and CD's that have been manufactured. We play sports with equipment that has been manufactured. We play with games or toys that started as raw materials and were soon converted into useful products for us to enjoy. Our lives would be much different if it weren't for industries that produce useful and convenient items.

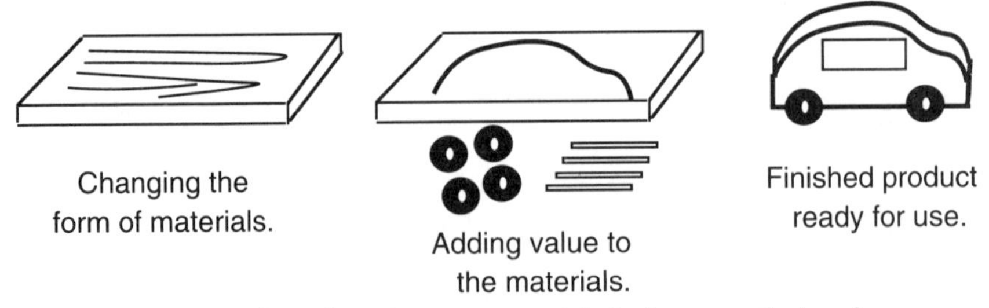


*Figure 3-1: Manufacturing changes materials to increase their value.*

Young learners must contemplate what their lives would be like without manufactured products.

By taking a birds-eye look at manufacturing, we see that manufactured products provide an easier and more comfortable lifestyle for all of us. Also, manufacturing firms provide a source of income for millions of managers and workers so that they can support themselves and a family. Products are shipped to stores in which other individuals are needed to transport the product and sell the items. In essence, manufacturing has given our country a great deal of capital, economic power, and stability. Manufacturing industry is a major influence in our society and it is important that elementary students understand how manufacturing affects their lives and society. After all, our young citizens will one day need to make rational decisions in their roles as producers or consumers.

This chapter will review the nature of manufacturing education at the elementary level. Specific sections will address the role of teachers, administrators, and young learners in developing an effective program. Numerous classroom and laboratory activities, many interdisciplinary in nature, will be discussed.

## VALUE OF STUDYING MANUFACTURING

Educators have an on-going task of preparing students to function effectively in an ever-changing technological world. The learning process starts when students are very young. Since a child's character is so moldable at this age, exploring the world of manufacturing can help students learn many principles about the modern workforce. For instance, positive attitudes toward work and desirable work ethics must be taught at the elementary level. Responsibility, dependability, and cooperation are among the many characteristics of an individual that can be nurtured in an elementary level manufacturing program.

Through the study of manufacturing, students explore concepts and tasks related to our industrial society. They begin to relate to the importance of manufacturing in their lives and learn that our society changes and grows rapidly with the discovery of new technologies. Today's manufacturers must be flexible and adapt to changes in technology in order to produce new and better products. An example of one modern topic that should be addressed involves automation. The need for a person to push levers to operate a machine is decreasing due to the development of automation. Further, because of advanced automation, educators have an even greater task of preparing students. With the "pushing levers" jobs disappearing from the workforce, there is little demand for low skilled individuals in either our educational process or local factories. Therefore, one of our challenges as educators is to produce researchers, designers, and engineers (i.e., those who can develop and maintain automation). This type of career planning obviously needs more attention as schools develop and implement new curriculum.

Constructional or "hands-on" activities can develop a dynamic framework for such a task of preparing our youth for a future in manufacturing. This hands-on approach to learning will be of great value as elementary educators attempt to prepare quality, intellectual, and competitive individuals, Figure 3-2.

Manufacturing at the elementary level should be an exploration experience designed to help students gain a basic understanding of how things are made and how manufacturing enterprises function. A primary goal in an exploration of manufacturing is to conduct hands-on, conceptual-based activities in order that students gain a knowledge and appreciation for this major segment of our industrial society. Students should be exposed to activities common to all manufacturing firms beginning with the designing of a product to the assembly line process. Instructional activities should focus on the tools, materials, processes, products, and occupations that exist in manufacturing industries. These concepts can easily be taught in conjunction with the existing elementary curriculum as outlined later in this chapter, Figure 3-3.

## HOW YOUNG STUDENTS LEARN

As elementary students begin to explore manufacturing, wonderful discoveries can be observed by the teacher and felt by the child. The development of personal skills and interpersonal relationships will become

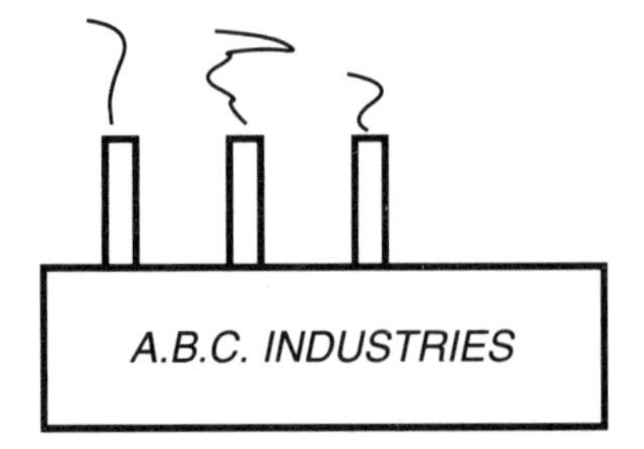


Industries manufacture products to meet consumer needs.

Society needs knowledgeable consumers and workers.

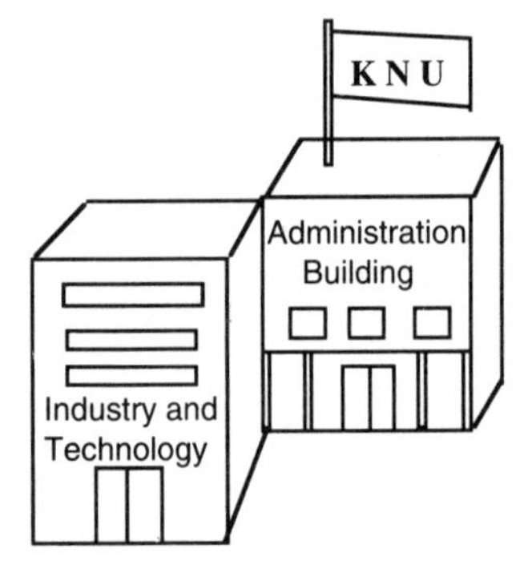


Education helps prepare people for society and the workplace.

A solid educational foundation is developed at the elementary level.

*Figure 3-2: Elementary manufacturing activities provide students with exploratory experiences related to living and working in a technological society.*

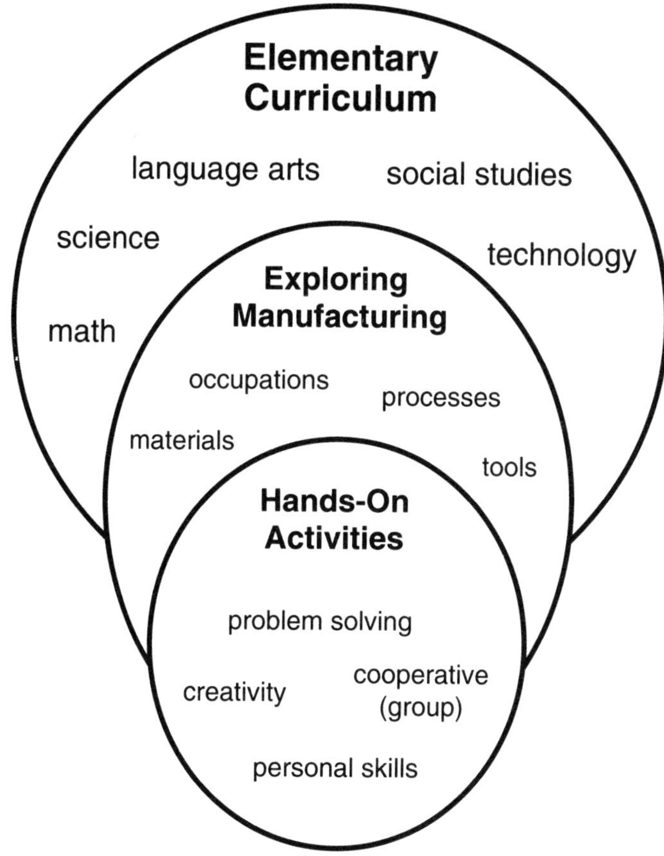


*Figure 3-3: The relationship of manufacturing experiences to the elementary curriculum.*

apparent. Integrating the study of technology and manufacturing can aid in a child's psychological (emotional), physical, and intellectual growth. This important development is linked to the cognitive, affective, and psychomotor domains of learning. At the elementary level it is very crucial to exercise great care in identifying, developing, and evaluating the domains in all young students.

## Cognitive Domain

In the cognitive domain, intellectual abilities and skills are developed. Elementary-aged children have a vivid imagination and a love for dramatic play. They are often unaware of their own limitations and their attention span is very short. Therefore, as educators, we must cultivate in children the ability to plan, be realistic, and foresee long-term outcomes and consequences.

The cognitive domain is made up of six major levels of learning from simple to the complex. The six levels and a short explanation follows:

| | |
|---|---|
| Knowledge: | Recalling the facts. |
| Comprehension: | Translating meanings into other forms (symbols, graphs, etc.). |
| Application: | Acting upon what has been comprehended. |
| Analysis: | Taking a closer look at principles, relationships, and techniques. |
| Synthesis: | Organizing and combining elements into a whole. |
| Evaluation: | Making judgements concerning worth and purpose. |

In a manufacturing program, all levels of the cognitive domain can be observed and/or developed. The chart in Figure 3-4 outlines several examples of developing intellectual skills in a manufacturing program.

## Affective Domain

The affective domain involves attitudes, interests, values, and appreciations. A student rarely acts in a given situation without having some feeling or attitude about the experience. Attitudes may develop without the student realizing it. Young children often begin to cry or pout easily without being able to understand their own feelings. A child's enthusiasm for a given activity may appear boundless for a moment, but quickly disappears in favor of another interest. These types of emotions are developmental characteristics found in the affective domain.

By nature, children are often selfish and somewhat boastful. A sense of ownership is established very early in a child and only as they mature will they begin to share. It is very important that children learn to interact and associate with the opposite sex, much like a working relationship in modern society. In the early years, boys and girls associate freely, but they tend to associate solely with members of their own sex beginning in grades 2 and 3. By the third grade the competitive spirit begins to be significant among the sexes. Students must learn that in today's workplace people work together on a daily basis. A sense of teamwork must be developed when doing group projects so that these early experiences of working together can be positive. The degree of a child's social devel-

| LEVEL | APPLICATION |
|---|---|
| Knowledge | * Identifying tools by name<br>* Recognizing that minerals come from the earth<br>* Realizing that manufacturing firms produce goods in a factory |
| Comprehension | * Understanding the flow of a production line<br>* Gaining information from a picture<br>* Understanding symbols used in production |
| Application | * Following directions on assembling a product<br>* Solving mathematical equations in calculating material needs<br>* Crosscutting a piece of material with the appropriate saw |
| Analysis | * Organizing a step-by-step procedure for producing a product<br>* Identifying the contributions of producers and consumers in society<br>* Identifying the sequence of processes in the making of a product |
| Synthesis | * Determining the demand of a product through a market survey<br>* Drawing conclusions from a simple experiment on material durability<br>* Identifying how a production line could be made more efficient |
| Evaluation | * Discussing the effects of manufacturing on individuals and society<br>* Determining advancements due to the manufacturing industry<br>* Identifying the impact of an innovative product |

*Figure 3-4: Examples of cognitive learning related to the study of manufacturing.*

opments can be greatly enhanced by the implementation of manufacturing activities.

Conceptual thinking and application is developed through the thought process, thus, it is largely cognitive in nature. However, since there is a direct relationship between conceptual development and attitude development, attention should be given in a classroom to subject matter which can be articulated and learned conceptually. When a subject matter is taught on a conceptual basis, a student can identify, understand, and explain their own feelings and attitudes about the subject. A manufacturing unit at the elementary level often encompasses the affective domain of learning. Concepts of manufacturing can be taught and experienced by students through hands-on/minds-on activities. These experiences should allow children to think, feel, and develop positive interests and appreciation for manufacturing and our technological society.

## Psychomotor Domain

The psychomotor domain involves the development of motor skills, or the ability to manipulate objects with coordination, dexterity, steadiness, and reaction time. Motor skills are simply the actions involving coordinated, muscular movement. The term psychomotor refers to the relationship between the mind and body.

At the elementary level, muscular development is a regulating factor in the selection of activities because it determines the nature of the tools and materials which students are able to successfully manipulate. In terms of human growth, the larger muscle groups are known to develop earlier. Those muscles include the limbs, shoulders, and back. The smaller muscles (fingers, eyes, etc.) develop more slowly. Therefore, the larger sweeping motions used to hammer and saw are easier for a small child long before the more exact control required to use screwdrivers and scissors. As a youngster's physical development progresses, greater degrees of precision are possible and more intricate tasks can be accomplished. Children's physical development is slow and gradual and little can be done to speed it up. Appropriate activities must be given with the learner's motor skills and abilities in mind. A child can become very discouraged if a task is too difficult due to their physical development. Brauchle (1985) notes that technology educators:

> . . . . want students to learn psychomotor skills because we believe that through the success experience of improving in specific skill areas, they will become better motivated and enabled to learn other things as well. We agree on three basic notions: (1) learning is enhanced by activity; (2) to engage in activity successfully one must develop a level of psychomotor skills; and (3) psychomotor skills can help develop an appreciation of tasks and processes. (p. 63)

In conclusion, exploring manufacturing at the elementary level can enhance a student's learning abilities as well as their social and physical skills. The development of relationships and being able to work together with peers is of great value and should begin in the elementary grades. Manufacturing activities can help students learn how they should function as a team as they share facilities, tools, assignments, and responsibilities. Students must also learn the importance of working together in an effective manner.

# IMPLEMENTING MANUFACTURING IN THE ELEMENTARY SETTING

The elementary curriculum consists of basic subjects such as language arts, mathematics, science, social studies, and technology. It is important to realize that the topic of manufacturing will be implemented into this interdisciplinary curriculum. Elementary educators have a rigid schedule and numerous subjects to cover in the course of a school year. Instructors are constantly changing their methods of teaching in order to maximize learning from their students. The implementation of any new content presents a further challenge for the elementary educator. This section will review suggestions for effectively implementing manufacturing topics.

To implement the study of manufacturing into an already established curriculum, the teacher will need to understand the different areas and aspects of manufacturing. Teachers must be motivated and enthusiastic about researching and creating activities, and possess the basic conviction that manufacturing is an important subject to begin exploring at the elementary level. Other criteria needed for implementation include motivated and innovative educators that have the future needs of their students clearly defined.

## Role of The Technology Teacher

Addressing technological topics at the elementary level have become increasingly popular due to the importance of technology in modern society. Yet introducing this new content is difficult for many elementary educators. For instance, it takes time to prepare activities and materials required for individual units. While it takes time, many teachers seek assistance from local industries or businesses regarding professional knowledge, donation of materials, or financial support. Elementary educators also attend workshops and conferences to learn of new industrial practices or management trends.

Today, there are four common ways of implementing technological units into the elementary setting:

Traveling Teacher: The teacher functions as a specialist moving from school to school and working directly with the classroom teacher.

School Specialist: A specialist functions much like an art or music teacher by coming into the classroom on a periodical basis.

Central Laboratory: This involves a teacher assigned to a dedicated laboratory within the school; students come to the room every so often for instructional activities.

Teacher Centered: The regular classroom teacher conducts the entire program within their home classroom.

In most cases, manufacturing content will be taught by the elementary educator in their home classroom. Therefore, the remainder of this chapter will include suggestions for a teacher-run program.

The typical elementary teacher already has a solid background in the traditional subjects of math, science, and language arts. They must become a generalist when it comes to technology, or specifically manufacturing, and must be able to combine several aspects of various fields of study. When assigning students to an activity they should supervise all actions going on in the classroom (many which may happen simultaneously). And finally, an elementary teacher is an acting resource person. Our society has become too complex for teachers to merely expound facts to children. Today, a teacher's role is to guide students to the appropriate texts, reference books, and supplementary materials available in the school. The teacher must often rely on other means than their own knowledge to aid student learning.

During class and laboratory time, the teacher should cover step-by-step instructions or demonstrate procedures before or during the activities. Most activities can be done in the classroom using the standard facilities, readily available materials, and simple tools. But one reminder, most lower elementary-aged children need guidance while undertaking a physical task.

In conclusion, the effectiveness of implementing a manufacturing program is typically dependent upon the individual classroom teacher. The teacher bears most of the responsibility for the implementation. The teacher is the one who knows the children's abilities and readiness. Therefore, the teacher can plan the level of activity and its timing in order that the students have a greater chance of succeeding. Advanced planning is also suggested to insure that the appropriate tools, materials, and work and storage facilities are available at the correct time and in the correct quantity. Finally, the teacher should also take the initiative to constantly supply data to the principal in order that they might have solid evidence on which to support program growth.

## Administrators & Curriculum Consultants

In gearing up for the implementation of manufacturing, the roles of administrators and consultants should be discussed. For instance, the principal is the key administrator of an elementary school. They are the ones to whom teachers look to for guidance and assistance in starting a new program. It is vital that the principal be informed of all meetings, ideas, and developments. Keeping administrators involved and informed will help them understand the importance of implementing manufacturing content in the curriculum. One strategy is to invite the principal to observe and even participate in manufacturing activities with the students. This will help them see how the program is benefitting the students intellectually, socially, and physically. Once the top administrators become believers in the program, their enthusiasm and leadership will encourage and motivate other teachers. Remember, the overall success of the program could lie in the hands of an individual administrator.

A consultant can also increase the effectiveness of the program. In some cases, the district will provide a professional in the field. In the case of a manufacturing program, a certified technology educator could act as a resource person and partner in the program. The consultant could be from an area university, high school program, nearby middle school, or an elementary school with a successful technology program. The consultant can guide, assist, and evaluate the implementation efforts. They might also be an excellent source for media and bulletin board materials. A consultant might meet once a week with the teacher or participate in the classroom activity with the students. In many cases, the consultant will often speak to the building administrators regarding tools, facilities, materials, and scheduling needs.

# MANUFACTURING OBJECTIVES IN AN ELEMENTARY PROGRAM

Some of the major objectives of studying manufacturing in the elementary classroom are to develop problem-solving skills, enhance critical thinking abilities, develop an understanding of manufacturing systems, and promote activity based learning. Many conceptually based activities also challenge the student's abilities to produce creative solutions to stated problems. Further, many activities provide career exploration for the students. This section will review several of the important goals of an elementary manufacturing program.

## Creative Problem-Solving Skills

Problem-solving skills and creative thinking are skills that need to be developed at an early age. When a student is able to relate to something concrete and tangible (like many of the interdisciplinary manufacturing activities), they should be able to observe the nature of the problem and the situation at hand. Solving a problem can be motivating to a student. Problems that have solutions which are meaningful and of real concern to the student catch the student's attention. In order for students to have opportunities to develop problem-solving skills, the teacher must provide activities which utilize concrete experiences. To merely provide the experiences is not enough. The teacher must also instruct the child in how to think through situations in order to effectively solve the problem.

There are numerous methods to solving technological problems. However, a general outline to solving problems is to:

1) Closely observe the situation,

2) Begin to analyze the situation which leads to problem identification,

3) Ideate (brainstorm) several solutions to the problem,

4) Select the best solution and experiment with it, and

5) Evaluate the solution and make readjustments if necessary.

Most classroom and laboratory activities will involve challenging problem situations. These five basic steps should be followed whether the problem is simple or complex. From the attaching of wheels on a toy car to the designing of a product to meet an intricate need for the consumer, both types of problems are solved in the same fashion. At the elementary level some problems will be best attacked by a group of students who can brainstorm or "bounce" ideas off of one another. Other problems will be best solved by one or two individuals who can claim its solution as their own.

While a student is developing problem-solving and critical thinking abilities, we must also allow for individual creativity to enhance the solution. A child's creativity must not be restricted due to the laws of physics or to perceptions within our society. Young learner's creative abilities peak around grades 3-5 and often get suppressed in the higher grades. We, as educators, must allow young learners to create and solve on their own. Our future as an industrious nation depends on design, problem solving, and creativity. Therefore, this creativity should not be stifled because it does not meet the teacher's standards or liking. We must allow creativity to happen.

## Critical Thinking

Critical thinking involves the mental outlook one takes in the world. Critical thinkers acknowledge the inadequacies of one's current answers to important questions. They show skepticism about knowledge and values and have an interrogating spirit. Critical thinkers are those who question what is given and probe to determine "the way it is."

Critical thinking requires seeking more information beyond the problem-solving level. Manufacturing activities provide the opportunities to the student to develop critical thinking. Manufacturing activities, for example, allow students to question whether or not the correct design, materials, or processes were used in the manufacture of a product.

## Conceptual Learning

The implementation of manufacturing should center around conceptual, interdisciplinary activities. The teaching of concepts through activities is a very effective way of learning. A concept in its simplest term means *idea*. So, when the teacher selects a manufacturing topic and teaches it through a hands-on activity, self-motivation and a higher degree of learning will result. Many values can be supported by a conceptual-based manufacturing program, too.

## Exploration of Future Careers

All learning is said to be for the future. To prepare a foundation for living in a technological world, attitudes and skills about work must be developed early. Careers provide the primary social force in modern society. Most people wait to decide (at some point) what they want to be or do. Some people identified their careers in the primary grades and gave careful thought and consideration to the types of skills they should focus on throughout their education. Other people just took the first job available. Those individuals who carefully choose their careers are usually happier than those who took what came along.

To tell an individual that a particular career exists in the area of manufacturing is not enough. Students are entitled to some evidence and actual involvement in what that job may require both physically and mentally. For instance, when mentioning a career related to marketing products, bring in a dynamic guest speaker such as a marketing representative from a toy manufacturer. This will reinforce that a particular job exists and what it entails. Then try letting the students role-play the steps in marketing (selling) a product.

Manufacturing can provide hands-on activities and group experiences necessary for motivating students and giving them true-to-life experiences. By actively exploring their career interests and capabilities, students will be better equipped for making future decisions about career choices. In addition, the students can focus their attention on required abilities, training, and characteristics needed for job success. Self-worth, knowledge of career opportunities, and school work become more meaningful to young learners at this stage when their attention is in tune with future work.

## Integrated Approach

An integrated manufacturing program can be implemented directly into an established curriculum and provides many opportunities for students to:

- Build, construct, and express themselves creatively. The child who builds success and achievement with something tangible will often develop confidence and self-realization.
- Identify and broaden the understanding of concepts through firsthand experiences. Bringing several of the senses into a given learning experience augments the effectiveness of learning.
- Apply knowledge in a natural and realistic setting. In manufacturing, the drawing, measuring, calculating, and designing all utilize facts and skills learned in other facets of the elementary curriculum.
- Develop an increased desire or motivation to learn. A child can be highly motivated to read, discuss, and write information that he/she can put to use immediately and can see some concrete result.
- Utilize problem-solving skills. Hands-on/minds-on activities give a realism to problem solving which is a major benefit for young students.
- Develop behaviors, attitudes, and appreciations that are not directly related to the activity at hand or the product being produced. A child will learn important lessons such as sticking to the project until it is finished, putting away tools and materials, accepting constructional feedback from peers and the teacher, and assuming responsibility for a task that is important to an entire group.

Finally, desirable work habits and positive attitudes towards work can be developed at the elementary level. Employability of individuals and their successes or failures in the work force depend a great deal on their attitudes toward the desire to work, dependability, loyalty, responsibility, respect, and cooperation. These characteristics can all be developed and nurtured through an effective manufacturing program.

# SAMPLE APPLICATIONS

Teaching manufacturing concepts at the elementary level should be done through activities which are interdisciplinary in nature using the appropriate instructional strategies and supporting the identified content. In other words, a manufacturing unit or activity should incorporate two or more other disciplines. Also, the concepts should be easily transferred from one discipline to another. For example, when teaching language arts, you could have the students develop their writing and communication skills by advertising a particular product in both written and oral form, Figure 3-5. Start the activity by giving each student an outline for the product they are to advertise. Other skills they might encounter include measuring their ad for the available posterboard, figuring the best price of the product depending on the proposed market, or planning a commercial to target a certain social group. Math, language arts, and social studies can all be integrated into in this single activity.

Other suggested activities are outlined in Figures 3-6 through 3-8. In all of the activity examples, age and ability of students must be taken into account. The suggested activities listed are for no individual grade level or age group and are not in a particular order.

An instructional strategy is *how* the teacher is going to use the teaching/learning process to help students gain the knowledge or experience. An example of instructional strategy might be that a sample activity will a) initially be an individual writing assignment, then b) move on to group interaction, and c) conclude with a presentation to the class. The teacher must remember their role is to guide students through the activity while

**LANGUAGE ARTS**

| Concept | Sample Activity |
|---|---|
| Practice writing skills (procedures) | Show the students a simple product that has several parts and have them list the order of assembly. |
| Practice with reading and understanding charts or graphs (visuals) | Show the students a sample of a flow process chart or operation process chart and have them write or explain its meaning. |
| Sharpening communication (reading, writing, and speaking) skills | Have the students select a product to advertise and have them write and produce a commercial (videotape the feature if possible). |

*Figure 3-5: Integrated language arts/manufacturing learning experiences.*

## SOCIAL STUDIES

| Concept | Sample Activity |
|---|---|
| Review the history involved with the Industrial Revolution | Using snap-together cars, have the students assemble the pieces as custom versus continuous line production (to illustrate the "new" concept of mass production in the late-1700s). |
| People as consumers of manufactured goods | Have the students make a poster display of products in the same general category (sporting goods, frozen dairy items, etc.) |

*Figure 3-6: Integrated social studies/manufacturing learning experiences.*

## MATHEMATICS

| Concept | Sample Activity |
|---|---|
| Measuring | Set up workstations in which the students measure and record different features on a product. |
| Mathematical problem solving including story problems | Make up story problems involving the producing, buying, or selling of materials or products. Use tangible items when possible. |
| Counting, sorting, and arranging in consecutive order or by quality, color, etc. | Have them classify and inventory a variety of similar shaped products (dowels, straws, etc.) by size, length, or color. |

*Figure 3-7: Integrated mathematics/manufacturing learning experiences.*

| SCIENCE | |
|---|---|
| **Concept** | **Sample Activity** |
| Understand how raw materials are used | Introduce the changing of a sample material into an industrial or consumer product (lumber, paper, etc.). |
| Studying the six simple machines | Have the students design and test a simple machine to sort popcorn, then identify which of the six simple machines are found in their device. |

*Figure 3-8: Integrated science/manufacturing learning experiences.*

exposing them to various learning experiences. As the teacher develops activities to support the study of manufacturing, they should keep in mind that the activity should be short in duration. Remembering that a young learner's attention span is very short and may frequently need new stimulation, a teacher should be creative in developing activities. Activities should not be limited to the classroom. The entire school and community should be viewed as the total classroom.

As an elementary teacher covers new concepts of manufacturing, various instructional strategies should be used. Various methods include group problem-solving activities, class discussions, individual and group presentations, role-playing, and hands-on activities. All of these methods meet the student's essential needs in some fashion. Using a variety of teaching methods helps alleviate boredom and enhances learning.

Also, as a classroom teacher, feedback and evaluation is important. Some feedback from the student is necessary to determine if goals and objectives were achieved. Feedback and evaluation can come from testing (both written and oral), class discussion, or activities involving physical skills.

# FACILITIES AND RESOURCES

The incorporation of manufacturing activities in the elementary classroom can be enhanced by the availability of some common tools, additional storage space, and a work area. In preparing to implement a new unit in manufacturing, it is important to keep the following thoughts in the forefront of the planning process:

- Activities will typically be conducted in the regular classroom,
- The instruction is often provided by the classroom teacher,
- Tools and supplies are often kept in the classroom and should be portable,
- Safety is critical when using hand tools, power equipment, etc.,
- Activities may run concurrently with the regular elementary curriculum, and
- The class time available to perform activities should be short.

The diagram in Figure 3-9 is but one example of how to prepare the classroom for manufacturing activities. This arrangement would be appropriate for all elementary grades but especially the upper grades (3-6). A few simple tools and a storage cabinet(s) for student work are extremely helpful. A portable work table for some simple hammering and student work should be conveniently located in the room.

Equipment and supplies for the elementary classroom will vary from school to school depending on financial support, nature of the activities, and numerous related factors. The most appropriate changes to the facility include acquiring tools, work benches or stations, storage cabinets, and additional materials/supplies. These important resources are described in greater detail below.

**Tools or a tool cart:** A portable tool cart that includes a small working surface on top is helpful. Options include storing tools under a portable workbench, inside a saw horse with enclosed sides, on peg board mounted on a wall, or behind a hinged-door cabinet. Regardless of how tool storage is provided, care must be taken so that obtaining the tools is safe for the students. Tools should not be stored too high for the students. Cabinets or portable stations should be well designed so they won't roll or tip. Some tools that should be on hand include coping saws, hammers, screwdrivers, hand drills, C-clamps, files, backsaws, try-squares, and rulers. NOTE: This is not a complete list.

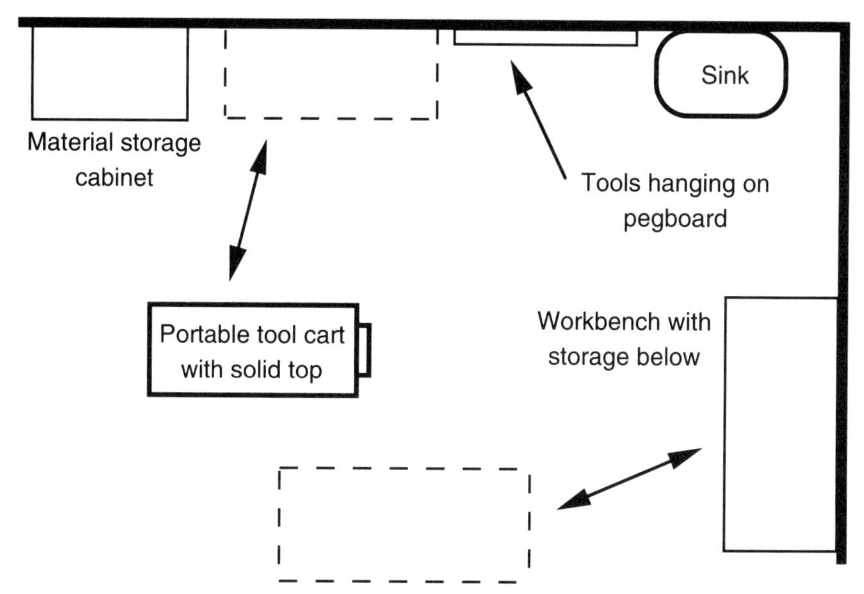


*Figure 3-9: Potential facility layout to support elementary manufacturing activities.*

**Workstations:** The main criteria for a workbench is that it is sturdy and the appropriate height for the children. For most activities, the student's desk or tabletops will provide the proper support. Workstations could be developed where the child would glue in one area, paint in another, and hammer at another station. A station can consist of three or more desks put together and covered with old newspapers. In Figure 3-9, note how a workbench can fit nicely against the wall and can also provide room below for tools or storage.

**Storage cabinets**: Most classrooms have commercial storage cabinets of various shapes and sizes. These same storage cabinets can be used to house materials and supplies strictly needed for manufacturing activities. Also, portable storage carts can be used. Portable carts which include shelving to hold plastic containers or tote trays are especially convenient. If the cart is topped with a heavy wood or formica surface, the tabletop can also serve as another workstation. With it being portable, it could also also be stored along a wall or moved to another room.

**Materials/Supplies:** Materials and supplies for a manufacturing program will vary from the lower to upper grades. Some general ideas of supplies and materials include the following:

| wood glue | sandpaper | scissors |
|---|---|---|
| pre-cut lumber | string | nails |
| styrofoam | cardboard | plastic |
| poster board | dowel rods | steel wool |
| coping saw blades | rubber gloves | screws |
| wood stain | latex paint | wire |
| paint brushes | plaster of paris | hardboard |
| rags | | |

It is important to plan well in advance for activities to be certain the correct supplies and materials are available. The teacher should bear in mind that many materials can be donated by parents, hardware stores, lumber yards, and other commercial outlets.

Instructional resources are also important to have on hand for the teacher's and student's use. These include technology-related magazines, textbooks, and articles about production topics. These items are also good for student research. The teacher may seek outside consultants for help identifying suggested reference books and magazine subscriptions that would be beneficial to the elementary students. A magazine and book rack is also an added feature to preparing the classroom facility.

Many excellent films and videotapes on manufacturing are available for elementary teachers. But the media must be interesting and short enough to hold the children's attention. Again, the teacher may want to seek the assistance of a consultant for suggested films or videotapes appropriate for the elementary level.

# SUMMARY

Elementary students need to recognize the importance of learning about manufacturing and should have positive experiences with manufacturing during grades 1-6. When students show enthusiasm for learning, the level of comprehension and retention increases. Therefore, it is important to structure the study of manufacturing with the student's abilities, needs, and interests in mind.

It is important the elementary manufacturing program focus on a broad range of objectives. Among the major objectives include developing an interdisciplinary experience that will address both individual and group problem-solving and critical thinking skills. Students should also be able to display creativity during their hands-on activities.

Most manufacturing programs, facilities, and resources at the elementary level are developed and/or conducted by the classroom teacher. The building principal should be involved in program planning and implementation. Outside consultants should be called upon to provide technical assistance, information concerning available media, and suggestions for facility and activity development. Perhaps most importantly, elementary teachers should seek advice from other educators who have already implemented manufacturing topics in their classrooms at either the elementary, middle school, or high school level.

# *REFERENCES*

Brauchle, P. E. (1985). Applications and implications of theoretical psychomotor learning models for industrial arts education. In J. M. Shemick (Ed.). (1974). *Industrial Arts for the Elementary School* (pp. 63-91). 34th Yearbook, American Council on Industrial Arts Teacher Education, Peoria, IL: McKnight Publishing Co.

Burke, J. (1978). *Connections*. Boston, MA: Little, Brown and Company.

Colder, C. R., Doyle, M. A., & Nannay, R. W. (1989). *Technology education for the elementary school*. Monograph 15, Reston, VA: Technology Education Children's Council.

DeVore, P. (1980). *Technology: An introduction*. Worcester, MA: Davis Publications.

Edmison, G. A., & Schwaller, A. E. (Eds.). (1992). *Delivery systems: Teaching strategies for technology education*. Reston, VA: International Technology Education Association.

Edmison, G. A., & Schwaller, A. E. (Eds.). (1992). *Approaches: Teaching strategies for technology education*. Reston, VA: International Technology Education Association.

Goetsch, D., & Nelson J. (1987). *Technology and you*. Albany, NY: Delmar Publishers, Inc.

Hacker, M., & Barden, R. (1987). *Living with technology*. Albany, NY: Delmar Publishers, Inc.

Hacker, M., & Barden, R. (1987). *Technology in your world*. Albany, NY: Delmar Publishers, Inc.

Jones, R. E., & Wright, J. R. (Eds.). (1986). *Implementing technology education*, 35th Yearbook, American Council on Industrial Arts Teacher Education, Bloomington, IL: McKnight Publishing Co.

McCrory, D., Todd, K., & Todd, R. (1985). *Understanding and using technology*. Worcester, MA: Davis Publications.

Miller, W. R., & Boyd, G. (1987). *Teaching children through constructional activities*. Urbana, IL: Griffon Press.

Savage, E., & Sterry, L. (Eds.). (1990). *A conceptual framework for technology education*. Reston, VA: International Technology Education Association.

Schobey, M. (1968). *Teaching children about technology*. Bloomington, IL: McKnight & McKnight.

Swierkos, M. L., & Morse, C. G. (1973). *Industrial arts in the elementary classroom.* Peoria, IL: Chas. A. Bennett Co., Inc.

Thrower, R. G., & Weber, R. D. (Eds.). (1974). *Industrial arts for the elementary school*, 23rd Yearbook, American Council on Industrial Arts Teacher Education, Bloomington, IL: McKnight Publishing Co.

Wright, T., Wright, P., Field, F., & Field, L. (1992). *How things are made.* Muncie, IN: Center for Implementing Technology Education, Ball State University.

CHAPTER 4

# Manufacturing Technology At The Middle School Level

Dr. Ray Shackelford (Professor)
*Ball State University, Muncie, IN*

and

Mr. Daniel Chapin (Technology Educator and Assistant Principal)
*Guion Creek Middle School, Indianapolis, IN*

The study of manufacturing in the middle school allows students to explore and gain an understanding of industrial manufacturing. This exploration into the field of manufacturing is one part of a basic technology education program. Along with the study of manufacturing, middle school students are also exposed to the study of other technologies.

Just as manufacturing is one area of exploration within the technology education curricula, the technology education program is one area of study within the general middle school curriculum. Current middle school philosophy is changing from an elective course offering structure similar to a "mini high school" to a required exploratory experience structure. This move affects the former elective areas such as technology education, home economics, and business.

This change in middle school philosophy is having a unique effect on manufacturing education, as well as technology education in general. Two distinct approaches to teaching manufacturing are developing as a result of this organizational change in the middle grades. One is for schools operating with elective courses on a semester basis and another for those schools that maintain required coursework for all students.

For schools that continue to operate as a junior high school offering elective courses on a semester basis, the model in Figure 4-1 describes a typical technology education program. In this scenario, cluster topics

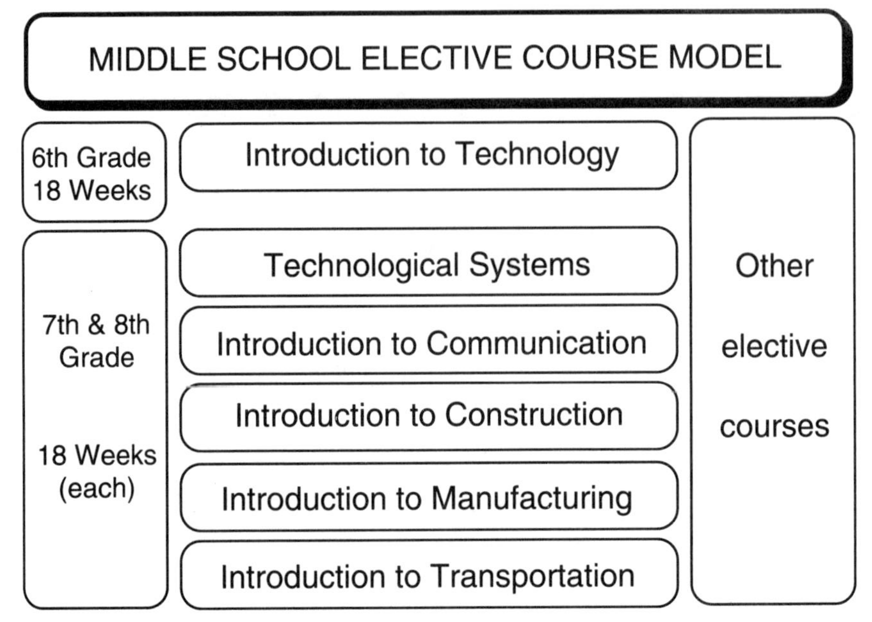


*Figure 4-1: Model of an elective course sequence.*

(such as manufacturing) would be an elective course. Students could select one of several technology-based classes during their middle school experience.

This sequence offers students the potential of one or more courses to choose from as electives offered by the technology education department. Students would also be selecting from other elective offerings available as well. This curriculum structure offers more depth in terms of content available to students. On the other hand, students are forced to choose between electives offered in the technology department and other elective courses in the school (music, art, business, etc.). With this elective approach, students have the opportunity to experience a thorough introduction to manufacturing technology.

Schools that utilize required exploratory courses as a basis for curriculum structure have made a choice not to offer depth in any one elective area. Instead they offer students breath in multiple elective areas. Middle school students explore all elective courses for an equal amount of time (in most cases) and do not have the option of choosing one curricular area over another. This allows for a thorough experience for every student; however, it limits the depth of understanding and experiences a student might gain in a particular area like manufacturing. For schools that operate on a required

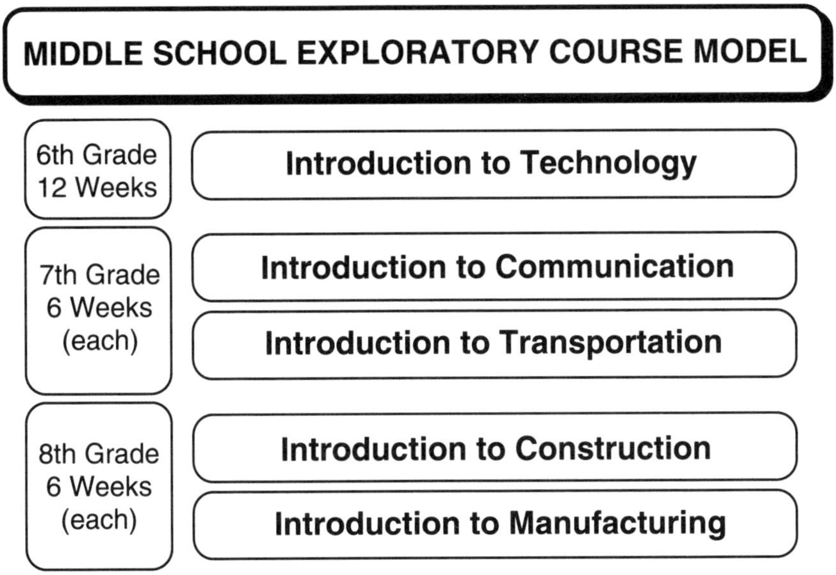

**MIDDLE SCHOOL EXPLORATORY COURSE MODEL**

|  |  |
|---|---|
| 6th Grade 12 Weeks | Introduction to Technology |
| 7th Grade 6 Weeks (each) | Introduction to Communication |
| | Introduction to Transportation |
| 8th Grade 6 Weeks (each) | Introduction to Construction |
| | Introduction to Manufacturing |

*Figure 4-2: Model of a required exploratory sequence.*

exploratory course format, a sample organizational structure is shown in Figure 4-2.

No matter which philosophy a middle grade school utilizes (elective semester courses or required exploratory courses) the focus on manufacturing education is the same; provide an introductory experience that is exciting for the student and encourages future study in the field of manufacturing.

## Benefits of the Middle School Manufacturing Program

Manufacturing is an area of technology education where students are able to apply the skills and content learned in other areas of technology education, as well as all other content areas in the school. Students have the opportunity to utilize math skills to solve "real world" problems when estimating the cost of a product. Students hone their communication skills when they present product ideas to their fellow classmates. These same students learn what is required of a manufacturing facility when they are asked to design a facility to meet the needs of a student-operated production line. Students apply the knowledge they learned in their science while testing manufacturing materials and evaluating the characteristics of these materials. Manufacturing education ties into many if not all areas within the school.

Just as the student's learning environment is reinforced through manufacturing education; so does manufacturing affect the student's learning environment. Students are bombarded with advertisements designed to persuade them to purchase a manufactured product. Most consumers (students included) blindly select and purchase manufactured products without knowing the processes required to design, produce, and market the items.

When a student completes a middle school manufacturing education program, he or she should be aware of manufacturing's relationships with all areas of technology. In addition, they should understanding the processes undertaken to convert raw materials into useful products to fulfill the consumer's desire or need.

# SAMPLE MIDDLE SCHOOL MANUFACTURING COURSE

This section outlines an introductory-level manufacturing course consisting of fundamental content, modules, learning experiences, instructional sequences, and interdisciplinary activities. The sample course will be called "Introduction to Manufacturing Technology" and would ideally serve as the lead-in course for a comprehensive program. It would support further coursework at the senior high level (Manufacturing Materials and Processes, Manufacturing Enterprise, etc.). The one semester course is designed for students in the 8th or 9th grades, but could be modified to fit any length or grade level. Content, activities, and the suggested time-line are all structured to support a beginning experience in manufacturing education.

## Introduction to Manufacturing Technology

Understanding manufacturing technologies and systems is important to an individual's technological literacy. All young citizens should understand how humans use machines and processes to convert materials, energy, and information to change the environment and meet human needs. All people depend on or use manufacturing, thus an understanding of manufacturing provides a basis for technological literacy and competence.

A course designed to introduce students to manufacturing technology and its related systems should include a study of four major themes: a) material processing, b) manufacturing management, c) product design, and d) the consequences of manufacturing. The "Introduction to Manufacturing Technology" course shown in Figure 4-3 includes modules and activities designed to help students develop understandings of these processes. As

## Introduction to Manufacturing Technology

**Course Description:** Introduction to Manufacturing Technology provides an introduction to manufacturing resources and systems. Students solve problems and make decisions necessary to (a) design and produce useful products, (b) use materials and processes, (c) operate a managed enterprise, and (d) investigate the consequences of manufacturing on individuals, society, and the environment.

**Course Outline**

| Module | Title and Activities | Days |
|---|---|---|
| 1 | Introduction to manufacturing | 5 |
| | Investigating manufacturing's role in society | |
| | Using manufacturing systems | |
| 2 | Manufacturing materials and processes | 10 |
| | Investigating and testing material properties | |
| | Evaluating material characteristics and report results | |
| | Producing industrial materials | |
| 3 | Parts manufacturing processes and concepts | 15 |
| | Investigating casting and molding processes | |
| | Examining forming processes | |
| | Performing separating processes | |
| 4 | Parts and product completion processes and concepts | 8 |
| | Exploring conditioning and finishing processes | |
| | Implementing assembling processes | |
| 5 | Enterprise organizing, financing, and managing | 5 |
| | Forming and staffing a mock enterprise | |
| | Raising capital and maintaining financial records | |
| 6 | Researching and developing products | 10 |
| | Establishing product criteria and sketching | |
| 7 | Manufacturing products | 17 |
| | Selecting and sequencing operations | |
| | Developing tooling | |
| | Establishing quality control and safety programs | |
| | Using computers to help develop, operate and control production systems | |
| | Conducting pilot and production runs | |
| 8 | Marketing (advertising and selling) products | 7 |
| 9 | Closing the enterprise | 3 |
| 10 | Analyzing consequences | 5 |

*Figure 4-3: Potential introductory manufacturing course (adapted from Indiana curriculum guide entitled Introduction to Manufacturing Technology, 1989).*

proposed, the course would be a teacher-directed course (e.g., most activities and experiences determined or directed by the teacher) with individual, small group, and full class activities. Because the course includes content related to materials science and management functions, it would strongly lend itself to interdisciplinary activities.

An introductory manufacturing course should be a hands-on/minds-on course involving students in exciting, interesting activities designed to stimulate and develop problem-solving and decision-making skills. Objectives for this course should foster the student's understanding of:

1. Manufacturing's influence in how people live, work, communicate, and move from one location to another.
2. A systems approach to the study of manufacturing technology.
3. The types, characteristics, and properties of engineered materials.
4. The managed activities used to design, engineer, manufacture, and market a product.
5. How manufacturing systems are designed and operated to produce products to meet human needs and wants.
6. The economic, environmental, and social consequences of modern manufacturing systems.

As suggested, a course designed to introduce students to manufacturing systems would include a study of: (a) material processing, (b) management, and (c) design. Material processing is often subdivided into manufacturing materials and processes.

*Manufacturing Materials.* Traditionally science has grouped materials into three forms—solids, liquids, and gases. The focus in manufacturing programs is almost entirely on solid materials. Solid materials are commonly referred to as engineered materials and grouped into material families such as metallics, polymers, ceramics, and composites, Figure 4-4. In an introductory course, students need to participate in learning experiences that help them understand material properties plus procedures for assessing, evaluating, and reporting these properties. The study also focuses on the methods and standards for determining and selecting appropriate materials for given tasks and the consequences of those decisions.

As students study and evaluate engineering materials, they should conduct tests and solve sample problems using destructive and nondestructive testing and inspection processes. These activities should help students enhance their understanding of different material properties (mechanical, physical, etc.). Common experiments might include tests to

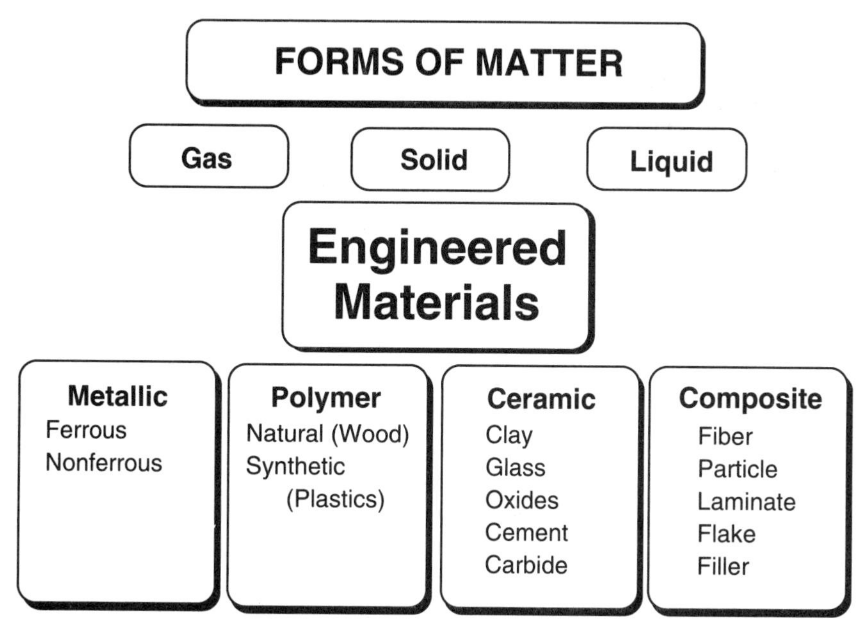


*Figure 4-4: Types of engineered materials to be investigated.*

investigate material mechanical properties (tensile, compression, impact, fatigue, hardness, bend, etc.). Further tests might involve physical characteristics including ductility, thermal conductivity, electrical properties, moisture content, chemical resistance, optical properties, and absorptivity.

During their study of materials, students also need to improve their understanding of the material's atomic structure. Further, the content should explore the types of material bonding (covalent, ionic, etc.). This provides an excellent opportunity for team teaching with colleagues from other disciplines and cooperative activities with the math or science programs.

From an industrial perspective, students should learn that materials can be produced to exhibit different characteristics and properties. They need to learn why different materials behave in different ways. They should also know how manufacturing technology can change materials and how they behave when external forces and conditions act upon them. More importantly, students should start to develop understandings of why certain materials are used for particular functions, under specific conditions, and in given situations. This can be readily accomplished using design and problem-solving activities. Perhaps the most familiar activities involve egg drop, bridge building, and solar collector themes.

***Manufacturing Processes.*** Processing is the second broad category of manufacturing materials and processes to be investigated in an introductory course. Students should be able to recognize how materials are (a) obtained from natural resources, (b) converted into industrial materials using primary processes, and (c) transformed into finished products using secondary processes.

Primary processes remove a material from its natural state and involve its conversion into an engineered material. Examples of primary operations include the reduction of iron ore into iron, timber into lumber, and synthetic chemicals into plastics. Such operations supply the raw materials used in manufacturing industries. To cover this content, teachers should design learning activities that permit students to use thermal, mechanical, and/or chemical processes to convert raw materials into familiar products (e.g., making plywood, particleboard, or bricks).

Although primary processes are important, students studying material processing should spend most of their time investigating secondary processes. Secondary processes are used to convert engineered materials (standard stock) into parts, subassemblies, or finished products, Figure 4-5.

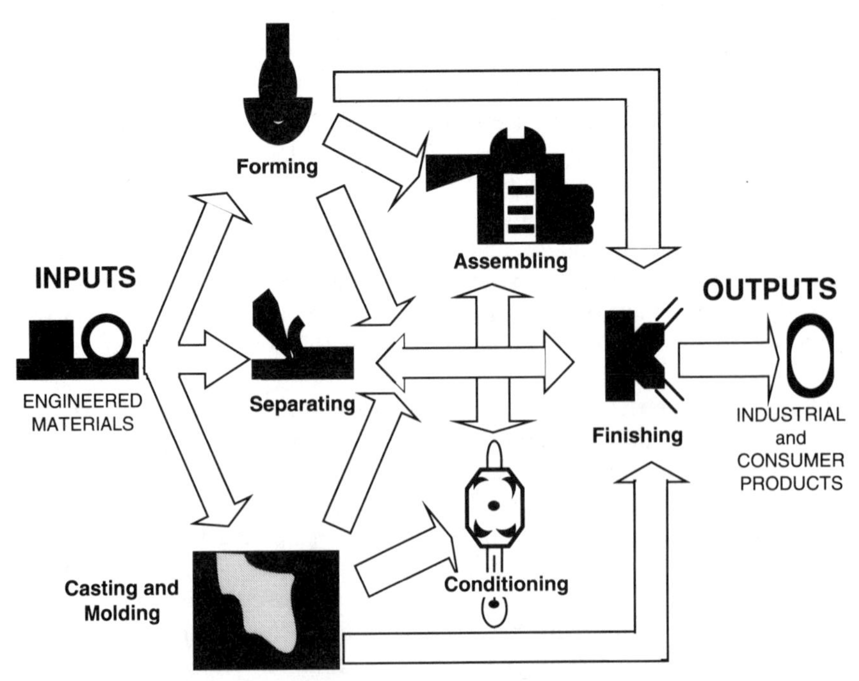


*Figure 4-5: Systems model of secondary manufacturing processes.*

Examples of secondary processes include milling, case-hardening, drop forging, investment casting, sawing, and riveting. To simplify the study of secondary processes, many manufacturing teachers organize these processes into the following six categories: a) casting and molding, b) forming, c) separating, d) conditioning, e) assembling, and f) finishing. Speed, economics, material characteristics, and quality considerations are all taken into account in determining the processes to be used.

In summary, the materials and processing portion of an introductory course should center around the families of engineered materials, material properties, primary processes, and the six categories of secondary processes. When planning laboratory activities, teachers should carefully select activities that provide an appropriate balance between processing and manufacturing materials. Common assignments used to support the study of manufacturing include permanent and one-shot casting and molding; hot and cold forming; separating by machining, sawing, shearing, and non-traditional cutting; adhesion, cohesion, and mechanical assembling; thermal, mechanical, chemical, and electrical conditioning; and coat or conversion finishing.

*Management.* In an Indiana curriculum guide entitled "Introduction to Manufacturing Technology" (1987), it is noted that a successful manufacturing enterprise must efficiently coordinate its materials processing, management, and research and development efforts. Management coordinates all inputs (people, capital, materials, finance, information, and energy) of a manufacturing enterprise. Perhaps the most important concept that students learn about a manufacturing enterprise is that the major goal of a firm is to make a *profit*. The more efficient use of materials, processes, design, and management results in a better profit.

Students participating in an introductory manufacturing technology course should study and participate in a managed enterprise where a product suggestion becomes a completed set of products. This can be either a teacher- or student-directed enterprise. During this experience, students should assume different levels of authority such as vice-presidents, plant managers, and production workers. In these roles they would role-play tasks associated with research and development, production, marketing, industrial relations, and financial affairs, Figure 4-6.

The operation of an enterprise experience provides several opportunities for group and interdisciplinary activities. For instance, to initiate the enterprise students must incorporate by developing the Articles of Incorporation, a charter, and by-laws. In addition, working capital is obtained and employees hired and trained. These tasks allow for cooperative efforts and interdisciplinary experiences between the business, economics, English, mathematics, and technology programs. It also provides a medium for

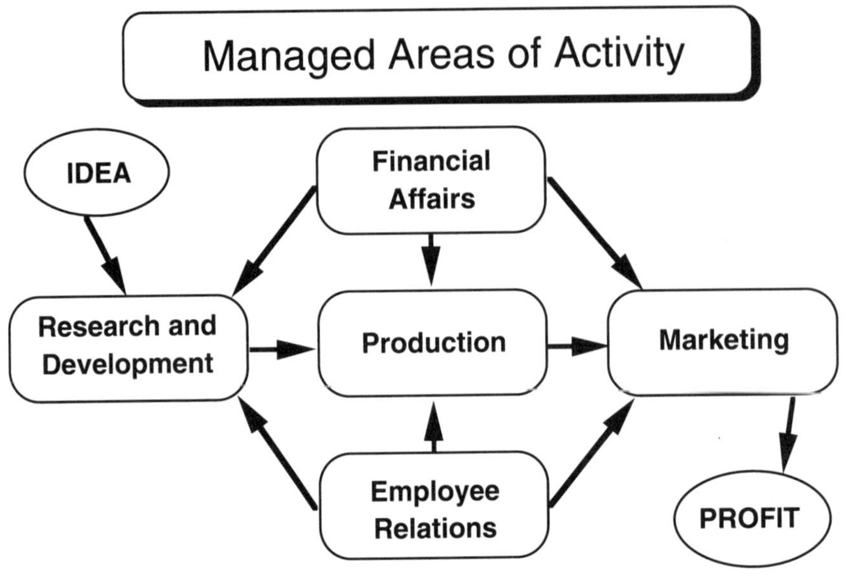


*Figure 4-6: The five managed activity areas.*

students to apply and transfer knowledge developed in other classes and life experiences.

Ideally, the enterprise experience would involve human and financial aspects of modern manufacturing. Students might be expected to apply for their jobs on the production line and be financially rewarded for their efforts. If school guidelines restrict direct payment of students, the profit from the enterprise might be used for a pizza party or to support transportation costs for a manufacturing field trip. The marketing department should prepare an appropriate award for the top sales representative in the class. Again, each of these tasks are important managed activities in a manufacturing firm.

*Product Design.* Before manufacturers can produce and successfully market a product, they must design an item that fulfills a consumer or industrial demand. Students can use this same design approach in developing product ideas during their material processing and enterprise experience. For example, students can follow a structured sequence when developing product suggestions for their enterprise, Figure 4-7. Ultimately, they could prepare an advertising program to convince perspective consumers there is a need for the product. In contrast, students could use a market research approach to generate product ideas by studying the needs and wants of potential consumers. Then they could design and manufacture a

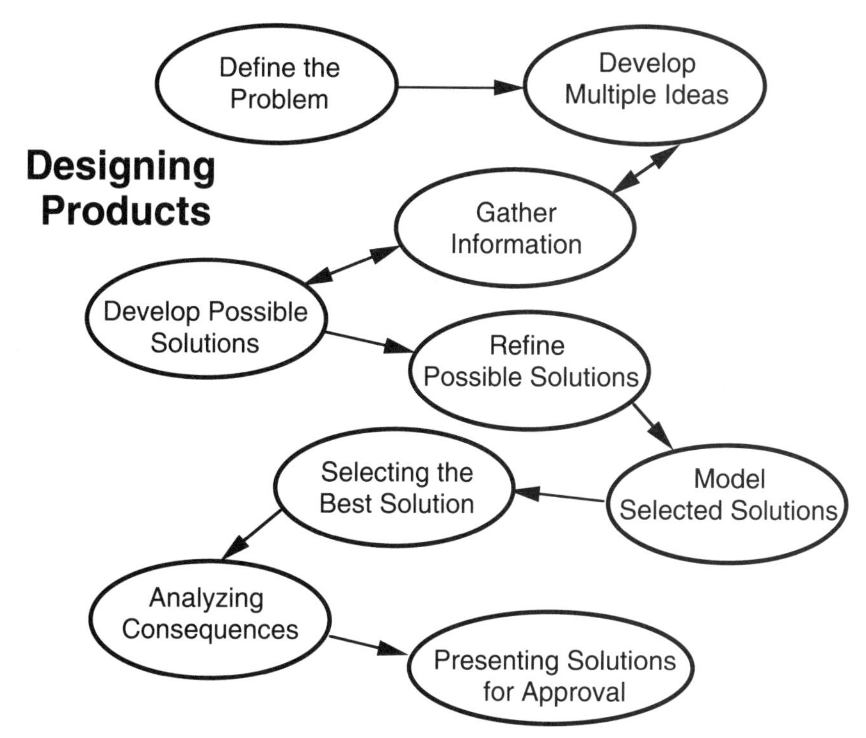


*Figure 4-7: Model of the product design process.*

product that follows the results of their market surveys. In either case, students are learning how the design process is used in modern manufacturing.

When students design products for a middle school enterprise, they must keep three key factors in mind. Their efforts should reflect the following:

- They must design for function (noting purpose, operation, and safety).
- They must be able to build it (reflecting on available resources, skills, and materials).
- They must be able to market it (keeping in mind the appearance and value).

To eliminate potential problems, teachers are encouraged to establish design criteria or parameters before the students begin designing products. These parameters should include costs, capabilities of the facility, and a rigid time frame. Qualities such as pride, quality of work, and being responsible for their efforts as well as the work of others should also be stressed.

Finally, it is recommended that each student participate in the design tasks. Each student should develop and present their own ideas for a product to be manufactured. From these suggestions, approximately three to five might be selected for further development and presented for final selection. The single, best product from the entire class might be approved for production. The selected product is then "released" to the research and development department where final detailed specifications are prepared including engineering drawings, a bill of materials, and specification sheets.

*Consequences of Manufacturing Products and Systems.* The study of modern technology should include a review of the positive and negative consequences of its devices and systems. In the area of manufacturing, this would involve a look into the economic, societal, and environmental impacts of products and production systems. For instance, personal and corporate tax income from the manufacturing sector represents a significant portion of local and state revenues (an obvious benefit for municipalities, schools, etc.). In contrast, unemployment statistics and news of plant closings often dominate the headlines, thus giving citizens a dim view of local industries and economies. These are two examples of the countless positive and negative impacts of manufacturing systems.

During manufacturing classroom and laboratory experiences, the teacher should work with students to identify local, regional, and international consequences of specific products and systems. A good example of such an activity would include the thorough analysis that should take place during the product design process when comparing multiple designs. Another time to consider the consequences of manufacturing systems is during the pilot run and after the final production run. This is an ideal time to discuss with the students what should be done with the scrap, waste, and remaining inventory. Suggestions might include recycling or using the leftover stock for other products.

## IMPLEMENTATION STRATEGIES

Any introductory course is going to be challenging. It embodies content from other courses, systems and processes to be studied in the same cluster. Yet, for many of the students, it is their first technology course. Thus, teachers may want to seek answers to the following three questions:

1. Since this course may be a student's only exposure to the study of manufacturing, what kind of a course should I provide?

2. Since this course may be the primary tool for recruiting students (male and female) into other courses in the manufacturing sequence, how do I make this course an exciting, meaningful, learning experience?
3. What can I do to insure that this course develops skills for future learning and encourages the transfer and application of knowledge to other disciplines and life experiences?

This section will review several key resources and instructional strategies to help implement introductory experiences in manufacturing. The suggestions are directed towards the middle school educator who is implementing manufacturing topics regardless of course title or time frame.

## Strategies and Activities

When discussing a course similar to the introductory manufacturing course described in Figure 4-3, many teachers ask "what do the kids do?" or "what are the activities?." This is a valid question, but tends to focus the discussion on implementation strategies (individual versus group activities, problem solving, role playing, etc.) and instructional activities and not the content. When implementing a new curriculum, teachers must remember that the strategies and activities support the content—*they are not the content*. With this in mind, classroom teachers have several resources for activities and suggestions for implementation strategies at their disposal. These include commercial vendors, publishers, other teachers, professional associations, curriculum guides, and the community.

Many times a new technology-based program must be implemented with limited financial support. This tends to limit the amount of materials and equipment that can be purchased for the manufacturing program. Although these limitations are often like asking a business education teacher to implement a new word-processing curriculum with manual typewriters, there are several resources available to help stretch the departmental budget. One example, in the area of material processing, is a series of activities developed and distributed by "The Parke System" (805 S. Devonshire, Springfield, MO 65802). Parke's activities are designed to support the student's exploration and understanding of secondary processes. Activities focus on processes such as a) permanent mold casting and molding, b) hot and cold forming, c) shearing and machining, d) organic and inorganic finishing, e) mechanical assembly, f) adhesion and cohesion bonding, g) heat treating and firing, and h) curing processes. The activities allow students to use an industrial process to produce a quality manufactured product. Activities are supported by student friendly materials that document each step of the manufacturing procedures.

Perhaps the best ideas for implementation strategies and activities come from fellow manufacturing teachers. After all, these activities have already been proven effective in classroom and laboratory situations. To implement the new activities, the instructor simply needs to make a few adaptations to match available facilities and equipment and meet the needs of their students. Activity suggestions are readily available at local, state, and international technology education conferences and meetings.

Another source of extremely useful information that often goes unnoticed are the materials available from textbook publishers. This includes many of the common items familiar to teachers (but often not purchased or used) such as teacher guides, student workbooks, and laboratory worksheets. Many publishers are now marketing textbook or course kits (binders) with each of their books. These binders include everything from sample quizzes and tests to daily lesson plans. These binders also include student handouts and transparency masters that match the information presented in the textbooks.

Otto (1992) stated that another excellent source of manufacturing activities is the Center for Implementing Technology Education (CITE) at Ball State University (Department of Industry and Technology, Muncie, IN 47306). The Center markets action-based activities that have been written and tested by public school teachers and teacher educators. Reports from teachers using the activities indicate they are effective and merely need to be adapted to individual settings. The CITE activities range in length from one day to a week or more. Some of the activities are conducted by individuals, while others are group-oriented. Group activities are broken down so that each person has a responsibility to the activity's successful completion. This reinforces the importance of group effectiveness, communication, and adaptability.

As written, CITE activities are not designed for any particular grade level or course. The CITE manufacturing activities are available in the following categories: general manufacturing, materials, processes, and enterprise. Each activity includes sections entitled introduction, concepts presented, objectives, equipment and supplies, procedures, related concepts, and references. All activities include a data recording sheet, checklist, and/or follow-up worksheet to check for student understanding.

Another valuable resource for introductory manufacturing curriculum and laboratory activities are the curriculum guides published by state or local departments of education or professional associations. These guides have proven to be very valuable resources. Too often, teachers attempt to "reinvent the wheel" by starting curriculum development efforts from scratch. It is far easier to contact state departments of education and ask for copies of their materials. In addition, look in professional journals for information on implementation strategies and activities, and check out the advertisements for materials available from professional associations and centers for the study of technology.

Finally, look in your own backyard. The community in which you teach has boundless resources including other faculty, parents, local businesses and industries, community organizations, and public and governmental agencies. Become familiar with their operations and establish contact persons in each sector. These contacts can lead to guest speakers, field trips, free or low-cost materials and equipment, and other valuable resources. Your own community will probably be your best resource once you have clearly identified and communicated your manufacturing program's goals and needs.

## SUMMARY

Manufacturing is critically important to every citizen; it impacts all individuals and nations. The introductory unit and/or course in manufacturing technology at the middle school level might be the only experience young students have in this important area. Therefore, middle school students should be able to participate in learning activities that review manufacturing systems and processes in detail.

Every school has a different format for required versus elective coursework. Two models for implementing manufacturing courses were discussed. In addition, a complete introductory-level manufacturing course was outlined. This dedicated course would include the themes of a) manufacturing materials and processes, b) management, c) product design, and d) the consequences of manufacturing products and systems. The one-semester course could be modified to fit any local schedule.

Numerous resources are available to support the study of manufacturing at the middle school level. Area industries, textbook publishers, state and local departments of education, and professional associations are among the best sources for instructional materials.

# *REFERENCES*

Daiber, R., & Erekson, T. (1991). *Manufacturing technology today and tomorrow*. Mission Hills, CA: Glencoe/McGraw-Hill.

Durkin, J. (1984). *Technology lab 2000*. San Diego, CA: Trastech Systems—Division of Creative Learning Systems.

Fales, J., Sheets, E., Mervich, G., & Dinan, D. (1986). *Manufacturing a basic text*. Encino: Glencoe.

Indiana Industrial Technology Curriculum Committee (1989). *Introduction to manufacturing technology*. Indianapolis, IN: State Department of Public Instruction.

International Technology Education Association. (1988). *Technology education . . . the new basic*. Reston, VA: Author.

Komacek, S., Lawson, A., & Horton, A. (1990). *Manufacturing technology*. Albany: Delmar.

Otto, R. (1992, presentation). *Implementing manufacturing*. Indiana Industrial Technology Education In-Service, Valparaiso, IN.

Shackelford, R. L. & Wescott, J. (1991). *Manufacturing materials and processes*. Muncie, IN: Ball State University.

Shackelford, R., Wright, T., & Haynes, F. (1987). *Industrial technology education—A guide for facility planning*. Indianapolis, IN: Indiana Department of Education.

Waetjen, W. (1985). People and culture in our technological society. In *Technology education : A perspective on implementation*. Reston, VA: International Technology Education Association.

Wright, R. T. (1990). *Manufacturing systems*. South Holland: Goodheart-Willcox.

Wright, R. T. & Shackelford, R. L. (1990). *Manufacturing system design and engineering*. Muncie, IN: Ball State University.

Wright, R. T. & Shackelford, R. L. (1990). *Product design and engineering*. Muncie, IN: Ball State University.

Wright, R. T. & Sterry, L. (1983). *Industry and technology education: A guide for curriculum designers, implementors, and teachers*. San Marcos, TX: Technical Foundation of America.

CHAPTER 5

# Manufacturing Technology At The High School Level

Dr. Ray Shackelford (Professor)
*Ball State University, Muncie, IN*

and

Mr. Richard Otto (Technology Teacher)
*Valparaiso (IN) High School*

In chapter one, Wright notes how technology involves a body of knowledge and actions used to apply resources in designing, producing, and using products, structures, and systems to extend the human potential. By itself, technology can do nothing. However, technology managed and controlled by people can be adapted to meet human needs and desires. To help understand, use, and control technological systems, models have been developed to explain its processes and purposes. These models typically have four common elements including inputs, processes, outputs, and feedback, Figure 5-1. The purpose of this section is to a) explain the role of manufacturing as a technological system and its interaction with other technologies and b) describe the purpose for studying manufacturing technology in the high school.

The systems model shown in Figure 5-1 is appropriate to all technological phenomenon including the study of manufacturing in technology education programs.

## Studying Manufacturing Technology

Many consider manufacturing to be a primary technological system. Humans have been dependent upon consumer and industrial goods throughout history. Advances in designing and producing these goods have

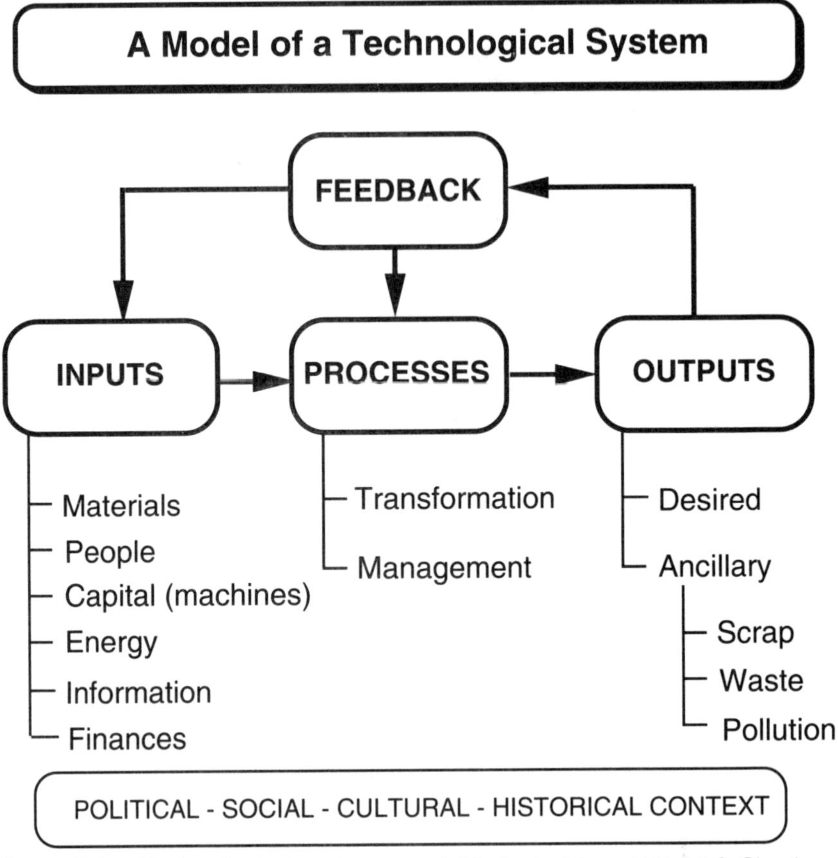


*Figure 5-1: Technological systems model (adapted from Wright & Shackelford, 1990a).*

led to materials, products, and forms of energy that transform the inputs of technology into messages, structures, manufactured goods, and the movement of objects. For instance, it is through manufacturing that we are able to monitor current events via television, radio, newspapers, computer networks, satellites, and magazines. Most information industries rely on manufacturing firms (publishers, computer firms, etc.) for their final products. The construction industry also uses manufactured tools and equipment to fabricate and install other manufactured materials such as nails, lumber, concrete and glass, windows, doors, kitchen fixtures, and furnaces. Furthermore, manufacturing is used to produce the vehicles that transport these goods and materials to the site.

The study of manufacturing technology should provide a foundation for technological literacy and competence. This study should provide students with the knowledge and skills to understand, use, and control manufacturing systems; comprehend its role in society and interrelationships with other areas and organizations; and recognize its effect on people, society, and the environment. Through the study of manufacturing technology, students should learn to prepare for and implement change, and make critical decisions regarding the future.

Manufacturing programs at the high school level should help students develop insights into how technological concepts, processes, and systems can be used to solve practical problems. The emphasis should be on the *doing* words such as creating, planning, problem solving, experimenting, and producing. Throughout the secondary experience, students should be involved in courses and laboratory activities that give them the opportunity to:

- Safely use tools,
- Apply technical concepts,
- Process materials, energy, and information,
- Design products and production systems,
- Review the consequences of manufacturing products and systems,
- Uncover and develop individual talents,
- Transfer and integrate knowledge learned in other subjects,
- Test management concepts and procedures,
- Sharpen critical thinking, problem-solving, and decision making skills, and
- Work in group situations to cooperatively address technological problems.

The following sections in this chapter will cover a potential model for implementing the study of manufacturing technology in the high school and suggest course structures and content. Each course in the recommended program will be fully described with course outlines, a module sequence, activities, etc. In addition, the discussion will include structures for implementing the manufacturing courses in small, medium, and large high schools. The recommended courses are followed by a review of trends and directions in manufacturing technology.

## Potential Curriculum Model for High School Programs

A proposed course structure for manufacturing was reported by Wright and Sterry (1983) in their work *Industry and Technology Education: A Guide for Curriculum Designers, Implementors, and Teachers*. Potential course titles in their model included: "Introduction to Industrial and Technical Systems," "Manufacturing Systems," "Manufacturing Materials and Processes," "Product and Production System Design," "Designing Products for Manufacture," "Manufacturing Production Systems," "Research and Development," and "Manufacturing Enterprise." The guide also included a manufacturing program for small, medium, and large school systems, Figure 5-2.

Manufacturing is often described as a human/technical adaptive system designed to efficiently utilize the inputs of technology to extract and convert materials into industrial standard stock and consumer products. While

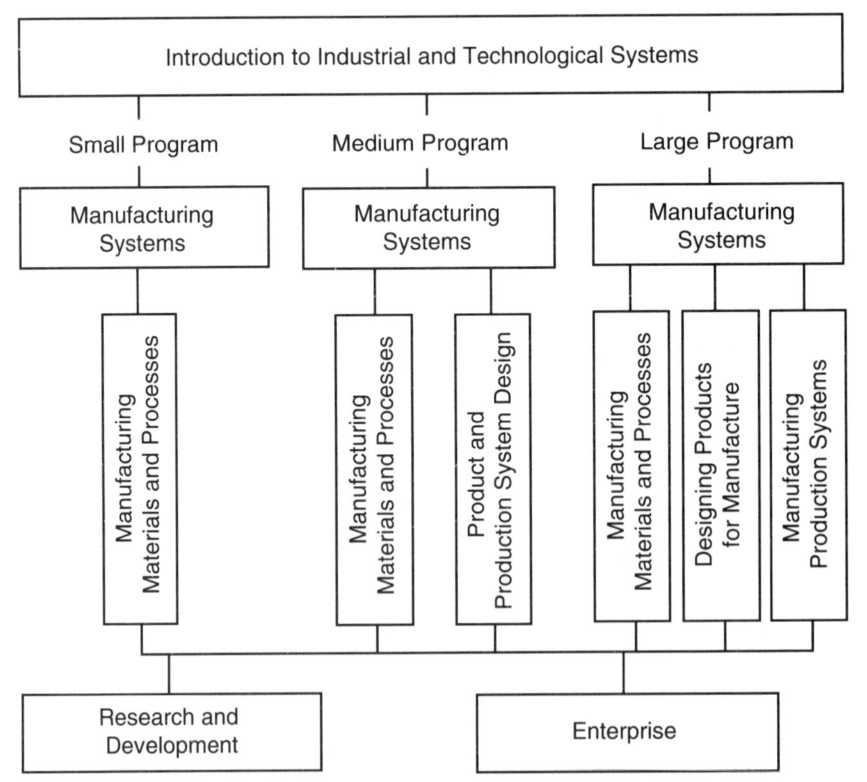


*Figure 5-2: Suggested manufacturing course sequence (adapted from Wright and Savage, 1983).*

suggesting that there may not be a perfect manufacturing curricula, Wright and Sterry (1983) did note that a complete broad-based curriculum would consist of three central themes including "a study of material processing, a study of managed production activities, and a study of the manufacturing enterprise" (p. 96).

Wright and Shackelford (1990) modified this model while developing curriculum materials for the Society of Manufacturing Engineers (SME). They suggested a five-course sequence for the study of manufacturing systems, Figure 5-3. Courses were proposed at three levels:

- An Introductory Course: Designed to study the basic systems used to convert materials into finished products.
- Specialization Courses: Designed to enhance the student's understanding of manufacturing materials and processes, and the design and engineering of products, and the development and engineering of production systems.
- A Synthesis Course: Designed to assist students to integrate and transfer learning experiences to new problems and decisions during the design, manufacture, and marketing of a product.

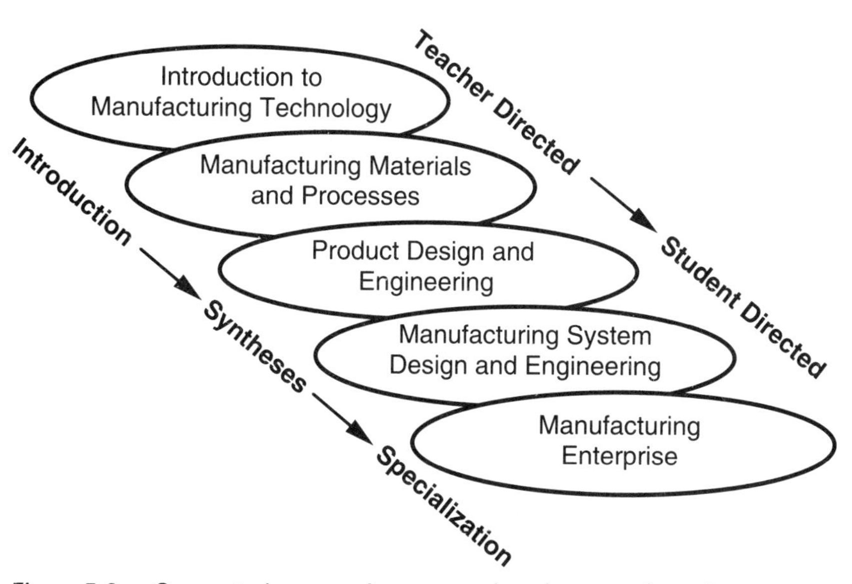


*Figure 5-3: Suggested courses in a comprehensive manufacturing program.*

Other curriculum developments often feature different levels and course titles, but the assumption is that most of the concepts, content, and underlying themes associated with these courses would likely appear in anyone's materials. In practice, the major differences between manufacturing practitioners, state and local curriculum guides, and publisher's materials appears in how the content is organized rather than a significant variation in concepts and content.

The proceeding section provides an outline of objectives and content for specific upper-level courses. Selected activities and interdisciplinary practices will also be described to help operationalize each course and summarize the planned student learning experiences.

# POTENTIAL COURSE STRUCTURES AND CONTENT FOR THE STUDY OF MANUFACTURING AT THE HIGH SCHOOL LEVEL

It would be impossible and inappropriate for high school students to study "all" there is to know about manufacturing because of time restrictions, the objectives of technology education, and the interest and motivation of high school students. Therefore, educators must determine what topics are pertinent to the needs of the student, society, and the body of knowledge to be studied. These topics should be subdivided into general themes (content) related to manufacturing technologies.

An introductory-level course should precede any advanced study at the secondary level. The "Introduction to Manufacturing Technology" course described in Chapter 4 would ideally serve this purpose. This basic course might be offered in the middle school or freshman grades. Then several courses would be designed to build upon the topics covered in both the elementary and middle school programs. Furthermore, there should be a natural progression from teacher-centered to student-centered learning as students take additional coursework. The advanced courses described in this section involve the following themes: manufacturing materials and processes, product design, manufacturing system design, and manufacturing enterprises.

## Manufacturing Materials and Processes

We, as a society, are dependent upon the materials and processes used to manufacture products that fulfill our basic needs. Thus, no study of manufacturing would be complete without a study of the materials and processes used to fabricate industrial and consumer goods. Processes involve tasks such as milling, sawing, heat treating, bending, stapling, and drilling; and familiar industrial materials like glass, copper, polystyrene, aluminum, cement, rubber, plywood, and carbide. However, a complete study of individual materials and processes would be difficult and endless.

Courses structured to concentrate on individual material conversion processes or materials (e.g., woods or machine shop) often fail to examine the many similarities that exist between modern materials and processing techniques. Unfortunately, these classes also place an emphasis on the learning of facts and isolated skills; the learning becomes slow, tedious, tiresome, and many times nontransferable. A more constructive method for studying manufacturing materials and processes is to concentrate on their concepts, principles, properties, families, and/or classifications. The sample course illustrated in Figure 5-4 is designed to support the understanding of current processes and materials and the development of learning tools to guide future understandings.

The goal of this one semester course is to introduce a) the properties of engineered materials, b) the manufacturing of standard stock, (c) the processing of engineering materials into products, and (d) the problem-solving, decision making, and organizational skills needed to investigate and effectively use industrial materials and processes. To achieve this goal, teachers and students need to work cooperatively to address five objectives:

1. To develop and reinforce the understanding of the properties and characteristics of engineered materials and manufacturing processes through action-based, laboratory-centered learning experiences.

2. To develop conceptual frameworks that aid the investigation, application, and transfer of current and future manufacturing materials and processes.

3. To develop the student's ability to observe, explore, problem solve, and make critical decisions related to manufacturing materials and processes.

4. To apply interdisciplinary concepts and learning to the investigation and understanding of manufacturing materials and processes.

5. To understand how manufacturing materials and processes are used to complete products to meet human needs and wants, and the consequences of those actions.

Although all forms of materials (solid, liquid, and gas) are used in the manufacturing of products, many practitioners focus their study of materials

**Manufacturing Materials and Processes**

**Course Description:** Manufacturing Materials and Processes provides students an opportunity to study manufacturing materials and how these materials are processed into parts and products. Students in this course solve problems and make decisions related to manufacturing materials and processes through hands-on activities used to test materials and transform standard stock into finished products.

**Course Outline**

| Module | Title and Activities | Days |
|---|---|---|
| 1 | Introduction to manufacturing materials and processes<br>Using materials to solve problems<br>Manufacturing products | 5 |
| 2 | Types of manufacturing materials<br>Investigating and analyzing manufacturing materials<br>Producing standard stock | 5 |
| 3 | Testing and evaluating material properties<br>Performing destructive and nondestructive tests<br>Using the computer to evaluate material properties and report results | 14 |
| 4 | Casting and molding processes<br>Investigating casting and molding principles<br>Performing and evaluating casting and molding processes | 9 |
| 5 | Forming processes<br>Examining forming principles<br>Performing and evaluating forming processes | 8 |
| 6 | Separating processes<br>Exploring separating principles<br>Performing and evaluating separating processes | 15 |
| 7 | Conditioning processes<br>Investigating conditioning principles<br>Performing and evaluating conditioning processes | 3 |
| 8 | Assembling processes<br>Exploring assembling principles<br>Performing and evaluating assembling processes | 7 |
| 9 | Finishing processes<br>Examining finishing principles<br>Performing and evaluating finishing processes | 5 |
| 10 | Designing and analyzing processes and materials | 5 |
| 11 | Analyzing consequences to the individual, society, and the environment | 5 |

*Figure 5-4: Sample course in Manufacturing Materials and Processes with suggested modules, activities, and time frame (adapted from Shackelford & Wescott, 1991).*

on the solid (or what are commonly referred to as engineered) materials. Modern industrial materials can be grouped into four families: 1) polymers, 2) metallics, 3) ceramics, and 4) composites. The study of these materials involves exploring the types of materials and their properties. The characteristics of different materials are determined by their internal structure. Each structure exhibits a certain set of properties which can be classified as physical, mechanical, chemical, thermal, electrical-magnetic, acoustical, optical, etc., Figure 5-5.

As students explore various resources, they need to be actively involved in materials testing (both destructive and nondestructive) and inspection assignments. Common testing operations involve tensile, hardness, torsion, compression, and shear tests. Students should perform and collect data which permit them to analyze the properties of metallic, polymers, ceramic, and composite materials and products. During inspection operations, students should examine materials and products to locate flaws, defects, or other undesirable qualities. Inspection might involve a review with ultrasonic or radiographic equipment. During all testing and inspection operations, students should become familiar with testing variables, specimen

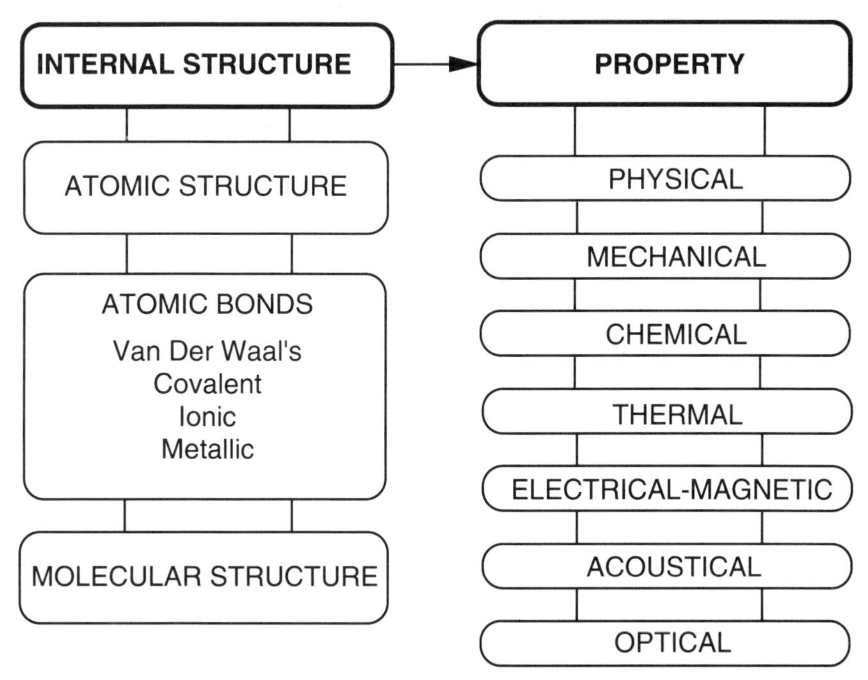


*Figure 5-5: Material structures and properties.*

characteristics, specialized equipment, testing conditions and procedures, testing economy, and the accuracy required.

Each of these laboratory procedures result in observations, data, and findings. A primary course objective is that students understand and report their information to others. Ball (1989) suggests sharing and discussing activities with the English, math, and science teachers in the school. Results of technical investigations should be communicated effectively and clearly. This is often a challenge since the data involves unfamiliar characteristics such as proportional limits, yield points, specific gravity, ultimate strengths, moisture content, breaking strengths, percentages of elongation, conductivity, and alloy composition. Sharing and discussing activities like these can lead to interdisciplinary opportunities and experiences. As Rye (1989) points out, "We don't live in a world of little boxes . . . and we shouldn't be teaching in them either. Everything we do should utilize as many other areas of education and society as possible" (p. 33).

A course involving manufacturing processes should emphasize the important role of primary and secondary processing in the manufacture of industrial and consumer goods. Primary processes use thermal, mechanical, and chemical means to convert raw materials into industrial (standard) stock, Figure 5-6. Raw materials are produced using mining, harvesting, and drilling techniques. A typical primary process would include using raw materials like iron ore, limestone, and coke to produce pig iron. The pig iron could then be further processed into steel sheets, bars and structural shapes. Other examples include producing aluminum from bauxite, distilling gasoline and kerosene from petroleum, and sawing logs into lumber.

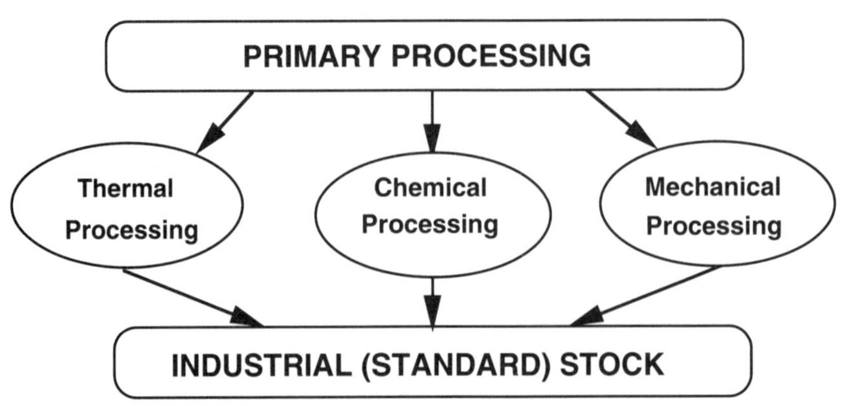


*Figure 5-6: Primary manufacturing processes used to produce standard stock.*

As suggested earlier, it would be impractical to cover all processing techniques in a single high school course. Many are too sophisticated for this level, but that still leaves hundreds to be examined. Fortunately, all manufacturing processes fall into six broad categories: casting and molding, forming, separating, conditioning, assembling, and finishing, Figure 5-7. These categories can enhance student learning, retention, and transfer in two ways. First, they simplify the tasks of describing and classifying various processes. Second, and perhaps more importantly, the corresponding concepts and principles allow for the study of future processing technologies.

According to Shackelford and Todd (1989), understanding materials and processes is easier when manufacturing processes are addressed as a *system*. Following a systems approach, material conversion involves four distinct

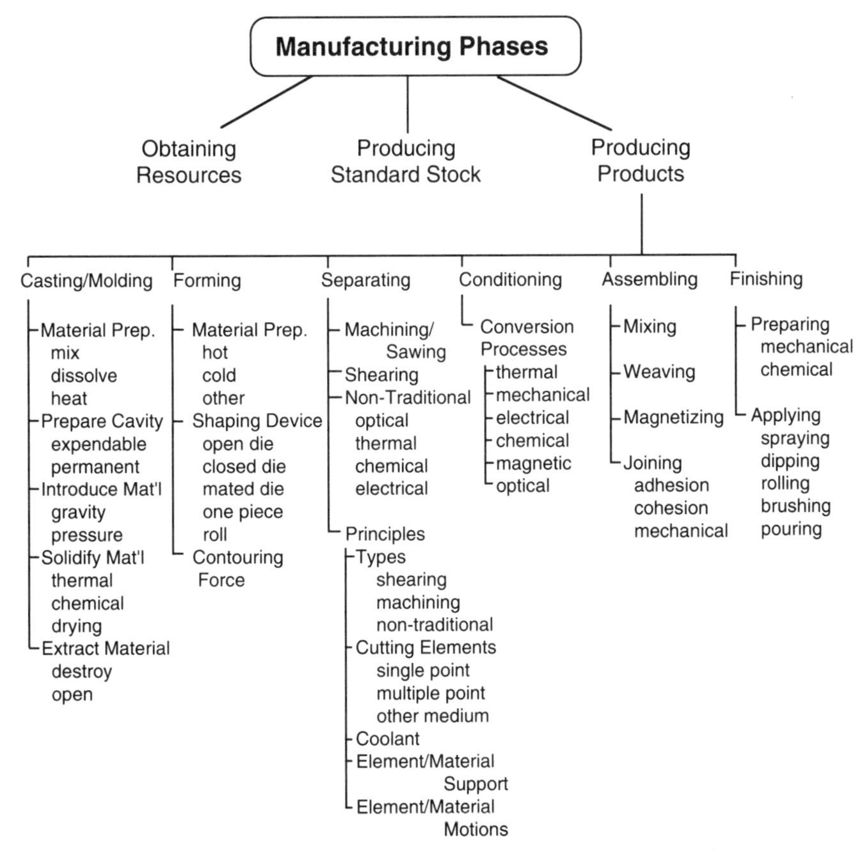


*Figure 5-7: Secondary manufacturing processes used to produce consumer and industrial products.*

elements: a) inputs in the form of raw or standard stock materials, b) the processing (or conversion) of materials, (c) the outputs of the system (products, materials, etc.), and (d) feedback in the form of process evaluation, assessment, or control. Common inputs of a material conversion system include materials, energy, and information. These interact with other inputs such as machinery and people to produce an output. Outputs of the system result in changing the shape, composition, and/or combination of materials.

As students perform classroom and laboratory activities, they should be challenged to investigate and report information about the industrial processes. Their reports might include a) a description of the processes, b) important environmental, social, and economic consequences, c) a description of the material(s) being processed, and d) a description of the processing techniques and principles used. For instance, a description of modern casting involves material preparation, cavity preparation, material introduction, material solidification, and material extraction. It is important students practice applying and documenting the appropriate processing principles during their laboratory work.

Materials and processing activities can be performed individually or in small groups. The activities can range from teacher-directed with limited alternatives and solutions (e.g., determining tensile strength) to student-initiated assignments with dramatic alternatives (e.g., determining the impacts of heavy metals being dumped into the water supply). Whether the activity in question is a seemingly simple injection molding experience or investigating the consequences of biodegradable versus non-biodegradable products, the teacher determines which activity will help the students better understand contemporary materials and processes.

## Product Design and Engineering

In today's global marketplace, manufacturing companies face tough competition. It is difficult to develop new products as well as improve older "dying" product lines. Companies must continually introduce new, innovative products that meet the needs and wants of customers, then design manufacturing systems to efficiently manufacture the products following increasing tougher quality standards.

In many technology education programs, students rarely have the opportunity to study the processes and systems used to develop, engineer, specify, and evaluate product suggestions. Too often, design courses are simply classes in mechanical or engineering drawing with a focus on orthographic projections and technical illustrations. Further, when multiple CAD systems become available, the same objective is often completed with computers rather than T-squares and drafting boards. Agreed, this may be an oversimplification, but the point is very few programs or courses focus on what should be the primary objective, "how are products designed for production?"

A sample course designed to meet this important objective, entitled Manufacturing Product Design and Engineering, is shown in Figure 5-8. The course was developed to help students learn how manufacturers identify and define a design problem, develop product ideas, engineer and specify products for manufacture, evaluate the proposed products, and present the optimum solution for approval. The emphasis here is on "doing" terms—identifying, problem solving, developing, evaluating, engineering, decision making, specifying, and presenting. The concept of design is addressed in a larger, industrial context with emphasis on designing products for mass production. This involves tasks such as problem identification and definition, market surveys, using interchangeable parts, cost analysis, designing for manufacture (DFM), creating mock-ups, and the preparation of engineering reports.

A product design and engineering course should be an action-based course with broad objectives such as:

1. Developing an understanding of how to best solve current and future product design problems.
2. Understanding and using various techniques to gather competitive data related to individual products and markets.
3. Functioning as an individual and/or member of a cooperative team to effectively and efficiently design, engineer, specify, and present a product for production.
4. Using techniques to model and communicate product ideas, materials, and specifications.
5. Applying mathematical, scientific, language, economic, business, and technological principles to design and engineer problems.
6. Efficiently designing and using products to meet human needs and wants, and making intelligent decisions concerning the consequences of those actions.

Students in a product design and engineering course often function as members of a design team. This allows students to assume the roles of actual designers. This also enhances realism and adds value to the activities and discussions. Challenging assignments, often in the form of design briefs, are used as an effective strategy to introduce the design process. During the

## Product Design and Engineering

**Course Description:** Product Design and Engineering provides students an opportunity to study the processes used to design, develop, and engineer products to meet human needs and wants. Students in this course solve problems and make decisions related to the tasks of designing, engineering, testing products for manufacture, and presenting ideas for managerial approval.

### Course Outline

| Module | Title and Activities | Days |
|---|---|---|
| 1 | Introduction to product design | 8 |
| | Solving problems | |
| | Making decisions | |
| | Designing vehicles and containers | |
| 2 | Identifying and defining design problems | 10 |
| | Working with design briefs | |
| | Defining problems | |
| | Gathering and interpreting data | |
| | Using word processing, spread sheets, and business graphics | |
| 3 | Developing and evaluating product ideas | 20 |
| | Brainstorming | |
| | Sketching rough ideas | |
| | Refining sketches | |
| | Fabricating mock-ups | |
| | Testing materials and products | |
| 4 | Engineering and specifying products | 37 |
| | Making detailed drawings | |
| | Preparing assembly drawings | |
| | Developing bill of materials | |
| | Performing cost analysis | |
| | Using CAD | |
| 5 | Analyzing consequences | 5 |
| | To the individual and society | |
| | To the environment | |
| 6 | Preparing ideas for managerial approval | 5 |
| | Developing engineering reports | |
| | Giving oral reports | |

*Figure 5-8: Sample course in Product Design and Engineering with suggested modules, activities, and time frame (adapted from Shackelford & Wescott, 1991).*

course, students should develop product designs by completing the following ten tasks:

| | |
|---|---|
| 1) Identify the problem: | Realizing a human need or want that can be met through a new or improved product. |
| 2) Defining the problem: | Developing a verbal or written description of the need, function, etc. the product must satisfy. |
| 3) Gathering information: | Acquiring background information and data that will be helpful in designing the product. |
| 4) Developing possible solutions: | Establishing several potential solutions to the defined problem (ideation/conceptualization). |
| 5) Refining possible solutions: | Selecting the best solutions and features from the conceptualization stage and further developing those ideas. |
| 6) Modeling selected solutions: | Fabricating physical, graphic, or mathematical models of the refined solutions. |
| 7) Selecting the best solution: | Analyzing solutions against defined statements of need, function, and/or criteria to identify which solution best fulfills the problem as defined. |
| 8) Interpreting the product: | Preparing, engineering, specifying, and communicating product designs through verbal, graphic, and mathematical means. |
| 9) Analyzing consequences: | Investigating and comparing potential economic, societal, and environmental consequences of the proposed solution. |
| 10) Seeking product approval: | Communicating the product (solution) to higher authority for approval. (Wright & Shackelford, 1990a) |

As students continue their study of the product design process, their focus on the important, early tasks of identifying the problem, defining the problem, and gathering information about the problem. Their product development efforts will likely incorporate one of the two popular strategies for developing consumer goods. They may tend to follow either the production approach or the market approach.

Using a production approach, companies develop products that will match their equipment, plant capacity, the skills of their workers, and present marketing systems. Most high-volume, low-cost consumer products are produced utilizing this approach. Several products fashioned using this approach are cosmetics, laundry detergents, and toothpaste. When manufacturers follow the market approach, they use market studies to determine the wants or needs of their clientele. In practice, the company is in a reactionary mode, striving to fulfill the desires of the customer. This approach can be implemented by either asking the customer to identify a need and producing the items to meet their need, or by developing a product and creating a demand for the product through aggressive advertising.

Fales, Sheets, Mervich, and Dinan (1986) recommend that students use small teams and brainstorming techniques to internally generate ideas during the design process. Students should clearly identify and define the problem, then focus on the prospective marketplace by gathering information relative to consumer demand. Social research (in the form of market surveys) are often used to gather such data. These surveys attempt to determine demographic, technical, and marketing information. A typical survey should result in a wide range of responses from people of different age groups, sexes, races, and ethnic backgrounds; various degrees of purchasing power; and, a mixture of professions/jobs.

Developing and conducting market surveys often provides an opportunity to integrate other disciplines. Using market data to make informed decisions includes mathematical, business, and research skills. Survey instruments must be designed, word-processed, administered, tabulated, and placed in graph form. Next, the data is analyzed and used to revise and clarify product definition, function, and criteria.

The next phase of the design process involves product development. This task requires that ideas be generated, refined, and improved. Student efforts would include developing alternative solutions, refining possible solutions, modeling the most promising solutions, and selecting the optimal solution. Product development builds upon the knowledge base (e.g., problem definition and consumer demand) by exploring and communicating potential product solutions. The product development phase uses sketching, drawing, and model building as tools to communicate proposed ideas to others, Figure 5-9.

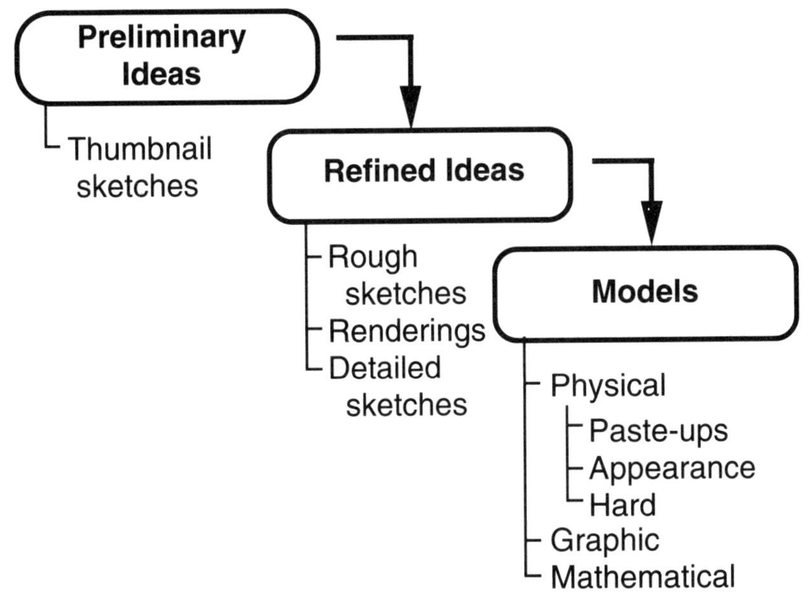


*Figure 5-9: Means of communicating ideas during the product development process.*

Product design involves generating dozens of thumbnail sketches to quickly capture preliminary ideas. These sketches often foster new ideas and relationships. Eventually, the ideas leading to a series of decisions wherein the optimal ideas are selected for further development. Rough sketches permit designers to add more detail to the preliminary designs. The greater detail enhances the exchange of ideas with other members of the design team. Detailed sketches are prepared from the most promising ideas (sketches). Renderings are high quality sketches showing shadows, textures, color, etc. Ultimately, models are completed from all these design sources.

With detail sketches, renderings, and models in hand, students typically schedule a product planning meeting to gain tentative approval of individual features and designs. During this session, preliminary ideas are presented and discussed with higher authorities (e.g., company management, team leaders, or the instructor). Key factors must be finalized including product appearance, function, safety, maintenance, cost, producibility, and marketability (Fales et al., 1986). The discussion might also review the limitations of the firm—knowledge and skill of workers, machine and tool limitations, space, time, and money.

Wright and Shackelford (1990b) point out that once students have received authorization for their product ideas, they are ready to prepare product specifications. Product specifications are used to communicate ideas and approved solutions to those who must manufacture the product. This information is forwarded to manufacturing engineers, marketing personnel, tooling and packaging designers, purchasing supervisors, quality control engineers, and plant management. In short, the design team must communicate to all manufacturing personnel the exact sizes, shape, configuration, and tolerances. This is accomplished through engineering drawing, specifications sheets, and bills of materials, Figure 5-10.

Students may use a CAD system to help prepare orthographic and assembly drawings of their approved solutions. Technical drawings should communicate the necessary dimensions and required tolerances for each part, subassembly, and final assembly. Specification sheets and bills of materials should also be prepared using computer word-processing and spreadsheet programs.

The culminating activity in this course involves students, as members of a product design team, presenting their developed suggestions for formal approval. This is different than the initial design presentation in that approval for the release of a specific product design is being sought. The presentation should include an oral presentation supported by a written report. A sample agenda for the presentation might include an introduction, description and definition of the problem, documentation of the idea's development, product specifications, an analysis of the product's costs, and a review of the consequences of the product on individuals, society, and the environment.

In conclusion, all companies face a common problem of continually developing new products to remain competitive. In the past, manufacturers could invent something new or unique, then enjoy a monopoly (and big

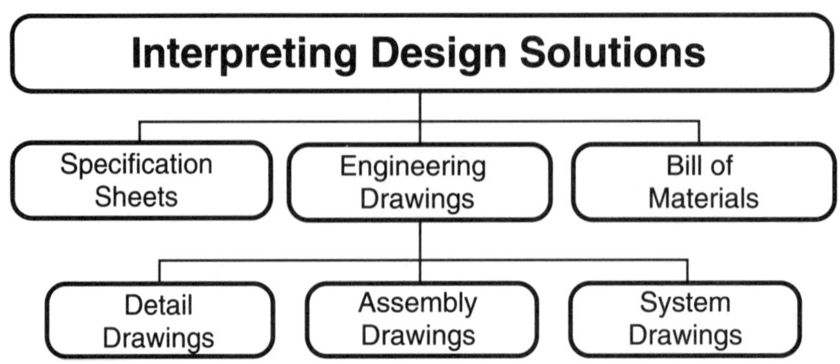


*Figure 5-10: Techniques to communicate product specifications (adapted from Wright and Shackelford, 1990a).*

profits) for generations. Today, the process of developing new products is endless as firms attempt to match fluctuating demands. While the design process may vary from company to company, the goal is the same—to design products which consumers want or need, when they want it, and with the quality they demand. The implementation of a product design and engineering course in a manufacturing sequence will give students the opportunity to study and use processes that help companies achieve this goal.

## Manufacturing System Design and Engineering

Once a product has been developed and approved for production, a manufacturing system must be designed, installed, and operated to efficiently produce it. During this phase a) manufacturing operations must be selected and sequenced; b) facilities engineered; c) tooling and production needs identified, designed, and fabricated; and d) control systems and programs established, Figure 5-11. Once these activities are completed, the manufacturing system must then be piloted, reviewed, and modifications made before full-scale production can begin.

It is important students be given the opportunity to study and use the systems and processes necessary to convert a designer's idea into a finished product. One method of accomplishing this objective is through a specialization course as shown in Figure 5-12. The primary goal of the "Manufacturing Systems Design and Engineering" course is to involve students in the study of different manufacturing systems (custom, intermittent, and continuous manufacturing) and production activities (manufacturing engineering, production planning and control, quality control, and manufacturing). The students would be challenged to design and engineer an efficient manufacturing system for an identified product. Young students completing this course should be able to:

1. Describe how production supports and integrates with other managed areas of a manufacturing firm.
2. Understand and use various systems and processes to develop an efficient production system.
3. Apply mathematical, scientific, language, business, and technological principles to design and engineer a manufacturing system.
4. Function as an individual and/or member of a team to effectively design and engineer a manufacturing system.
5. Pilot and operate a manufacturing system and analyze the system efficiency.

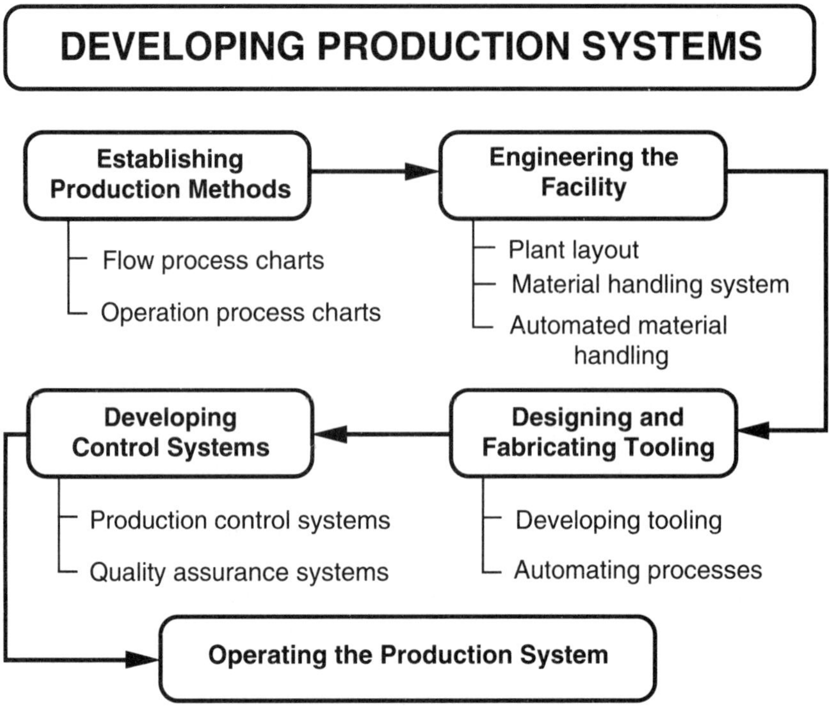

**DEVELOPING PRODUCTION SYSTEMS**

**Establishing Production Methods**
- Flow process charts
- Operation process charts

**Engineering the Facility**
- Plant layout
- Material handling system
- Automated material handling

**Designing and Fabricating Tooling**
- Developing tooling
- Automating processes

**Developing Control Systems**
- Production control systems
- Quality assurance systems

**Operating the Production System**

*Figure 5-11: Process of developing production systems (adapted from Wright and Shackelford, 1990b).*

6) Develop understandings and learning tools for solving current and future manufacturing system design and engineering problems.

7) Make informed decisions concerning the consequences of modern manufacturing systems.

The course illustrated in Figure 5-12 incorporates numerous "high tech" systems into the study of manufacturing. Computers, automated systems, CAD and spreadsheet software, and bar coding equipment are integrated into the program. This integration emphasizes their use, application, or role in modern manufacturing rather than a focus on particular pieces of equipment (CNC lathes, robots, etc.). However, the concept of integration must go beyond the use of things, processes, or systems and foster the application of math, science, language, and business principles. Opportunities for interdisciplinary activities include the following:

## Manufacturing System Design and Engineering

**Course Description:** Manufacturing System Design and Engineering provides students an opportunity to study the processes of designing, engineering, and testing systems used to manufacture products to meet human needs and wants. Students in this course solve problems and make decisions necessary to establish efficient production methods, engineer manufacturing facilities, design and fabricate tooling, develop quality and production control systems, and operate a manufacturing system to produce an identified product.

**Course Outline**

| Module | Title and Activities | Days |
|---|---|---|
| 1 | Introduction to manufacturing systems<br>Analyzing different manufacturing systems<br>Participating in custom vs continuous production lines | 6 |
| 2 | Establishing production methods<br>Developing operation sheets<br>Developing flow processes<br>Developing operation charts<br>Using computers to help develop production systems | 7 |
| 3 | Engineering manufacturing facilities<br>Using the computer to plan plant layouts<br>Designing and fabricating material handling systems<br>Programing robots | 15 |
| 4 | Using CAD to design and fabricate tooling<br>Designing tooling<br>Fabricating tooling<br>Programing CNC equipment | 28 |
| 5 | Developing control systems<br>Planning and fabricating quality control devices<br>Designing and inventory control systems<br>Using bar coding and spreadsheets | 10 |
| 6 | Operating manufacturing systems<br>Conducting pilot runs<br>Analyzing results<br>Modifying necessary operations, systems, and tooling | 10 |
| 7 | Analyzing consequences<br>To the individual and society<br>To the environment | 5 |

*Figure 5-12: Sample course in Manufacturing System Design and Engineering with suggested modules, activities, and time frame (adapted from Wright & Shackelford, 1990a).*

| | |
|---|---|
| Mathematics: | Time and motion studies, spreadsheet analysis, tolerances, inventory, cost control, laws of probability, speeds, quality control and sampling, production rates, wage incentives, work sampling, cycle time, and productivity. |
| Science: | Material selection and specifications, energy consumption, worker fatigue, material processing, speed, force, and ergonomics. |
| Language: | Flow charting, preparing operation and inspection charts, job descriptions, listing tooling requirement, quality control program materials, worker training, formal presentations, and bill of materials. |
| Business: | Lead time, just-in-time inventory systems, quality circles, operating costs, management, method engineering, and work measurement. |

A course in manufacturing system design and engineering encourages students to solve problems and make critical decisions necessary to develop an efficient production system. During the course introduction, students might participate in and analyze different production systems. This teacher-centered activity could use a simple learning experience to accomplish two major tasks. First, it would help students develop an understanding of different manufacturing systems (e.g., custom versus continuous systems). Secondly, it could serve to introduce future content such as workstation design, tooling, material handling, process flow, plant layout, and quality control.

Following the introductory activity, one or more products should be selected to serve as the focus of all systems design efforts. The teacher typically identifies the product(s) based on class size, facilities, individual skills, material costs, and time. However, it is strongly suggested the students participate in the selection process to gain some measure of ownership in the course. Students would then establish the production methods necessary to produce the identified product(s).

Establishing a productive system is a cooperative activity. Komacek, Lawson, and Horton (1990) point out that production staff use methods engineering, manufacturing engineering, and quality control engineering to enhance productivity while lowering production costs. Individuals from each of these departments should help with the developmental efforts. This recent trend is called "simultaneous engineering" in the manufacturing sector.

Komacek et al., (1990) suggests that students plan the necessary processes by developing charts that communicate a proposed sequence. These graphic materials include operation process charts, flow process charts, and operation sheets. Operation and flow process charts use standard graphic symbols developed by the American Society of Mechanical Engineers to identify operation, inspection, transportations, delay, and storage acts.

Operation process charts list, in order, the operations and inspections necessary to manufacture each part of the product. Students should use product drawings, specification sheets, bill of materials, and their understanding of manufacturing processes to develop the operation process charts. As students consider available resources (tools, machines, people, etc.), these charts should be constantly revised until the most efficient system is identified.

The development of charts, planning worksheets, and other documentation can be greatly enhanced with the power of a computer. Operation process charts and flow process charts can both be completed using a CAD or "draw" program. Operations sheets, which identify specific operations, machines, and worker directions, can be prepared using word-processing software. More advanced programs may be required if students want to use cut and paste functions to include plan views of workstations or tooling setups in their operation sheets.

Wright (1990) points out that selecting and sequencing operations are only part of planning a manufacturing system. The facility, methods of material handling, and tooling must be engineered. The process of engineering a facility involves the placement of machines and other equipment to enhance product flow. Specifying the plant layout involves assigning space within the facility. Again, computers with CAD or draw software should be used to facilitate the engineering process. Many teachers have their students produce a master drawing of their school's facilities including walls, windows, cabinets, permanent pieces of equipment, and utility lines. Students then draw all pieces of equipment, material handling devices, robots, etc. to the same scale as the facility. Once all drawings are placed in a file with a corresponding menu or legend, plant design involves simple cut, copy, paste, and rotate functions. Further, if the software has a layers function, rooms, pieces of equipment, services, material handling devices, flow arrows, and descriptors can be placed on different layers to enhance readability. This process utilizes the true power of microcomputers to full advantage; new documentation can be prepared quickly using previously prepared files and features.

Manufacturing is based on the concept of interchangeable parts since "like parts must fit any like product" (Wright, 1990, p. 267). This presents high school students with the real world challenge of designing tooling that increases speed, accuracy, and safety while maintaining specified tolerances and productivity standards. Tooling includes fixtures, jigs, patterns and templates, and special cutters and dies. All student-built tooling should follow general guidelines including: clamping pressure, work support, dimensional control, ease of setup, ease of operation, provisions for chips, and safety.

Contemporary jigs and fixtures are often designed and fabricated using standardized parts or components. Pre-cut bases, quick-action clamps, and locating pins are helpful in building authentic tooling. Many teachers have students use these components to speed up the process of designing and building their jigs and fixtures. This improves the learning process since the activity maintains a design focus rather than becoming too labor intensive.

As part of their study of manufacturing system design, Daiber and Erekson (1991) recommend that students investigate CNC, CAD, computer-aided manufacturing (CAM), computer-integrated manufacturing (CIM), robotics, and flexible manufacturing systems (FMS). They suggest that it is important for students to know how these modern systems are used in manufacturing. As in industry, students in a system design and engineering course should use automation to increase the efficiency of their manufacturing system and reduce production costs. Robots can be used by students to perform processes that are dull, dirty, or dangerous. CNC can be used to produce multiple parts with great accuracy.

However, well-designed facilities, processes, and tooling are not enough. Today's manufacturing systems must have effective control systems. These include both a) production control and b) quality assurance systems. Wright and Shackelford (1990a) suggest that students develop their production control systems to monitor work and inventory control. Work control programs manage the labor required to complete production tasks and inventory control programs manage the amount of materials used to manufacture the product. In some manufacturing programs, students use bar coding and other computer-assisted programs to maintain inventories of supplies, raw materials, purchased parts, work-in-process, scrap, and finished products.

People will only continue to buy a product as long as it meets quality standards. Thus, an aggressive quality assurance program should be an integral part of this course. The goal of the quality assurance program is to insure that manufactured products meet design specifications with the smallest number of rejects possible. In today's global marketplace, manufacturers are striving for *zero* defects. During this high school course, students should be urged to maintain high quality standards. To accomplish this goal, students must a) design products with quality in mind, b) understand and support quality efforts, and c) inspect and remove materials, parts, and products not to specification.

Quality assurance programs are often subdivided into two areas—motivational and inspection. Motivational programs are developed to encourage individuals to keep quality in mind and to promote quality during design and manufacturing efforts. With an effective motivational program in place, students should then develop an inspection program designed to identify and remove defective materials, parts, subassemblies, and finished products.

Developing a manufacturing system is an informative, yet challenging assignment. The best way to evaluate the developed system is to operate and assess the total system. This means the students should a) install the system, b) test each component of the system, c) train workers to perform individual tasks, d) test the total system, e) operate the system for a specified period of time (e.g., one hour), f) take any corrective actions required, and g) operate the production system. Obviously, accurate control systems (inventory, quality, etc.) should be in place and used. A culminating module would include a thorough analysis of the consequences of the system to individuals, society, and the environment.

It is important that technological-literate members of society understand the systems and processes used to manufacture the products they use every day. This action-based course would provide an adequate focus on the systems used to produce industrial and consumer products.

## Manufacturing Enterprise

The introductory section to "Production Technology" in the *Illinois Industrial Technology Orientation Curriculum Guide* (1989) discusses how people and society have always placed a high priority on the manufacturing of products. Manufactured products are designed to fulfill human needs and wants, improve the quality of life, and facilitate economic growth. To improve our standard of living and enhance economic growth and development, manufacturing enterprises must constantly undergo significant change. These changes make it possible for industrial enterprises to manufacture products with greater efficiency, better accuracy, and fewer people.

The inclusion of an enterprise course as a synthesis-level experience in the manufacturing sequence gives students the opportunity to investigate and understand how these changes occur. The course should be designed to provide an in-depth look at the concepts and systems covered in previous courses (refer to Figure 5-2 and 5-3). This includes an opportunity to integrate and apply material processing, design, and management concepts in an organized system. (Note: Because many of these classes have been discussed in the explanation of the specialization courses, only topics unique to the enterprise course will be discussed in this section.)

The sample enterprise course shown in Figure 5-13 would support the in-depth study of a student-directed, educational enterprise. Through group interaction, students could enhance their technological literacy by studying and participating in the managed functional areas of a mock corporation. This course should enhance the students' ability to make decisions regarding the uses and consequences of manufacturing and make wise choices as future consumers.

The goal of an enterprise course is to provide an action-based experience where students organize and operate a simple company. Teachers and students should work to complete objectives such as:

1. Describing the relationships and functions of research and development, production, marketing, human relations, and financial affairs in an enterprise.
2. Functioning as an individual and member of a cooperative group effort to form and operate an enterprise to design, engineer, manufacture, and market a product.
3. Applying mathematical, scientific, language, economic, business, and technological principles to material processing, design, and management problems.
4. Understanding the functional tasks of planning, organizing, directing, and controlling various activities.
5. Efficiently producing products to meet human needs and wants.
6. Assessing the importance and consequences of manufacturing firms to society, the economy, and the environment.

In their study of modern manufacturing enterprises, students will likely encounter the same challenges faced by other companies as they seek to cut costs, increase flexibility, reduce defects, increase customer satisfaction, and enhance their competitiveness. They should also learn the importance of working with fellow students (workers) plus managers, suppliers, customers, etc. and how management has changed its strategies for interacting with skilled workers, Figure 5-14.

Teaching manufacturing (and, in particular a manufacturing enterprise course) is not always easy, particularly in the beginning. Students should be allowed to make simple mistakes and learn from their individual and group

## Manufacturing Enterprise

**Course Description:** Manufacturing Enterprise provides students an opportunity to study the processes used to establish, operate, and close an enterprise. Students in this course solve problems and make decisions necessary to form a company; finance and control operations; design, engineer, produce, and market a product; and manage and run the company's activities.

**Course Outline**

| Module | Title and Activities | Days |
|---|---|---|
| 1 | Introduction to a manufacturing enterprise | 6 |
| | Determining what an enterprise is | |
| | Analyzing different types of enterprises & management | |
| | Participating in a production system | |
| 2 | Developing a product for manufacture | 24 |
| | Identifying product criteria and needs | |
| | Generating and evaluating product ideas | |
| | Determining product's market, financial and technical feasibility | |
| | Using CAD to specify and engineer products | |
| | Selecting products for manufacture | |
| 3 | Organizing the enterprise | 5 |
| | Incorporating the enterprise | |
| | Developing a managerial structure | |
| | Staffing the organization | |
| | Establishing goals | |
| | Determining deadlines and schedules | |
| 4 | Operating the enterprise | 40 |
| | Developing production systems | |
| | Developing marketing systems | |
| | Developing personnel systems | |
| | Developing financial control systems | |
| | Using computers to help develop, operate and control systems | |
| | Testing the production line | |
| | Producing products to stated specifications | |
| | Marketing products | |
| 5 | Closing the enterprise | 5 |
| | Dissolving the corporation | |
| | Liquidating assets and analyzing success | |
| 6 | Analyzing consequences | 5 |
| | To society and the environment | |

*Figure 5-13: Potential secondary-level enterprise course (adapted from Indiana curriculum guide entitled Manufacturing Enterprise, 1989).*

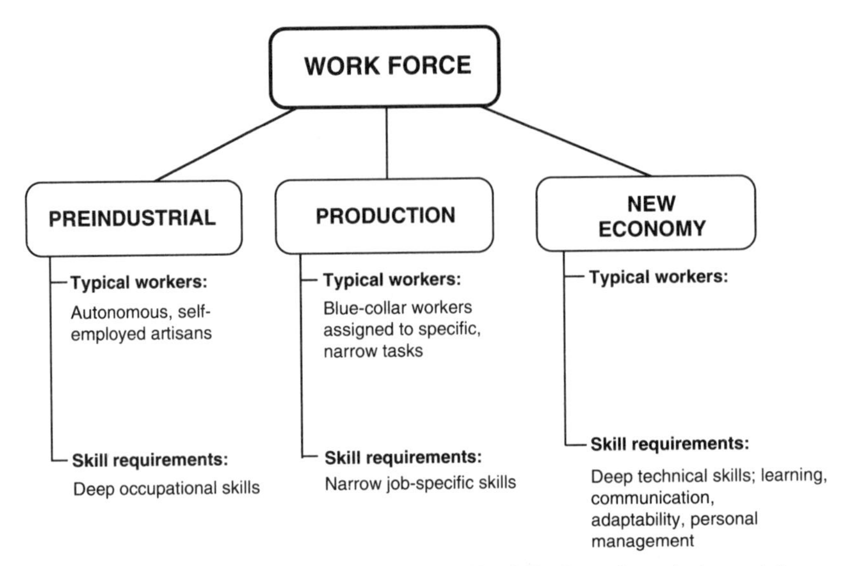


*Figure 5-14: Strategies for interacting with skilled workers (adapted from Tuma and Fischer, 1991).*

decisions. But students often find the increased responsibility difficult to grasp. Comments from Scott Beranek, a student from Rice Lake (WI) High School reveal some of the major challenges they faced and the changes in thinking and actions that occur. Beranek wrote in a Spring, 1986 *Manufacturing Forum* article entitled "The Student Based Enterprise: A Success Story" that when the students and teachers at Rice Lake developed their first enterprise, it was a unique experience for everyone. He observed:

> "It was difficult to describe the learning that took place in that first attempt to mass produce a product." (p. 3) A product (bulldozer) was selected that any of us could have easily produced as an individual. It was not a difficult product. "However it was very foreign for us to design a jig or fixture that would maintain a tolerance of 1⁄16 of an inch." (p. 3) Many of the jigs were crude at best and tended to work only when operated by its designer or builder. After further development and pilot testing "most of them" worked with different operators. We felt this was a major accomplishment. At the end of the scheduled production run " . . . we looked back at our accomplishments and all we saw were broken jigs and half finished products that only resembled the finished products. Sales did not go that well as we were at times embarrassed to take responsibility for creating such a product." (p. 3)

After the bulldozer experience Beranek wrote that the students wanted to redeem themselves by developing another product. They found that learning had taken place by both teachers and students. The previously complex task of producing jigs and fixtures was completed in only ten days, except for one student in the class. But Beranek explained that this envisioned problem turned out to be an excellent learning experience for all.

> The way it turned out he taught us the most significant lesson that we experienced in research and development. Three weeks in the making, his tooling debuted by processing 150 parts in less than 30 minutes. The point learned was that research and development pays off and has been a major part of all our successful enterprises since that day. (p. 4)

Quite often, our students have great insights—we simply need to step back and listen.

It is imperative that this course be student-centered. The instructor should play the role of consultant and resource person to react to ideas plus monitor plans and actions while not providing all the answers to routine problems. The instructor needs to keep an open mind and keep communication channels open to more than one idea or solution. Squier (1984) indicated that effective communication is an absolute must in a student-run enterprise. He found that the educational enterprise provides students with a true-to-life learning experience as they are exposed to non-traditional careers, problem solving, and cooperative learning.

Squier further stated that successful management techniques can improve motivation, teamwork, and productivity in students. For example, Douglas McGregor proposed two types of managers and labeled them "Manager X" and "Manager Y," Figure 5-15. Individuals following a "Manager X" style believe that people often dislike and avoid work and must be coerced or threatened to be productive, require very close supervision, and have little ambition. In contrast, "Manager Y" styles include people who are generally self-directed, willingly work towards common goals, and use imagination, ingenuity and creativity to solve problems.

Squier also reported that an enterprise experience is more positive when students understand their duties and responsibilities in the organization. A company personnel chart can help students visualize identified responsibilities and lines of communication and authority, Figure 5-16. Written job descriptions will also help clarify duties and identify expectations.

In a manufacturing enterprise students work as individuals and in small groups (departments, committees, etc.) to apply knowledge and skills to solve problems identified by the group. This requires successful cooperative

| McGregor's Theories X and Y | |
|---|---|
| Manager (Theory) X: Some traditional assumptions about people. | Manager (Theory) Y: Some modern assumptions about people. |
| 1. Most people dislike work, and they will avoid it when they can.<br>2. Most people must be coerced and threatened with punishment before they will work. They require close direction.<br>3. Most people prefer to be directed. They avoid responsibility and have little ambition. They are interested only in security. | 1. Work is a natural activity, like play or rest.<br>2. People are capable of self-direction and self-control if they are committed to objectives.<br>3. People will become committed to organizational objectives if they are rewarded for doing so.<br>4. The average person can learn both to accept and seek responsibility.<br>5. Many people in the general population have imagination, ingenuity, and creativity. |
| A teacher who believes these assumptions about students will probably not succeed in running an enterprise. | A teacher who cultivates these traits will probably succeed in managing and consulting with a student enterprise. |

*Figure 5-15: Outline of McGregor's management theories (adapted from Kreitner, 1983).*

efforts to reach desired outcomes. Many of the efforts can be enhanced by interdisciplinary activities. The implementation of a student enterprise is unique because of its almost limitlessness opportunities to incorporate other disciplines. It is simply a matter of time and creativity on an instructor's part as to how other areas can be integrated into the enterprise experience. A few examples include:

Speech: Developing presentations and running business meetings.

English: Producing product descriptions, market surveys, employee handbooks, job applications, etc.

Keyboarding: All computer usage.

Math: Product specifications, reporting tolerances, ordering materials, graphing financial data, etc.

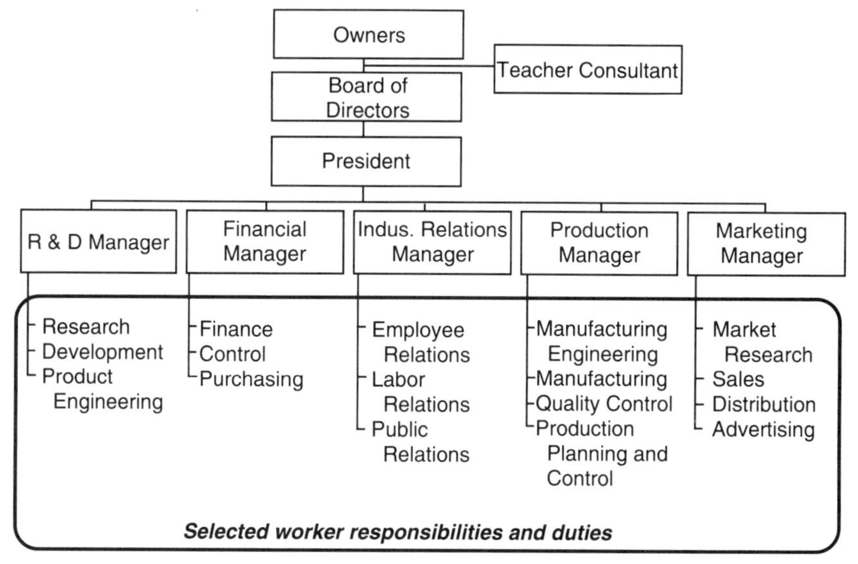


*Figure 5-16: Organizational chart of a student enterprise with potential worker responsibilities.*

Art: Advertising layouts and package designs.

Business: Spreadsheet applications, profit-and-loss statements, etc.

Geography: Impact studies.

As illustrated, many of the activities in the enterprise experience do not appear in previous manufacturing courses, modules, or units. For instance, this is the course in the manufacturing sequence to highlight the human (employee) relations, marketing, and financial affairs tasks. Students may assume such jobs as the corporate trainer, a purchasing agent, or the head of product distribution. Each duty has some relevance to the success of the student venture. Students quickly learn that all parts of the enterprise must work together to accomplish company goals.

Dr. W. Edwards Deming (1991) says if "businesses want to compete and survive in the new global market place, they must pursue the goal of constant improvement everywhere—from the assembly line to the board room" (p. 25). By implementing a dynamic, interdisciplinary, synthesis enterprise course, teachers can help students prepare for a life of constant change and improvement, and enhance their understanding of manufacturing and its role in the global society.

# RECOMMENDED STRUCTURES FOR IMPLEMENTING THE STUDY OF MANUFACTURING TECHNOLOGY IN SMALL, MEDIUM, AND LARGE SCHOOLS

Implementing a manufacturing technology program in the high school is dependent upon numerous elements. The critical items include the size of the school, demand for the program, availability and quality of instructional materials, number of technology education teachers, nature of the facilities, individual class sizes, budgets, equipment, and library resources. All these elements and resources must be taken into consideration whether the program is being implemented by a single teacher in a small school or by several teachers in a large high school.

This section will offer suggestions for structuring manufacturing programs in different settings. The suggestions will focus on adapting content, activities, and experiences rather than eliminating potential topics and activities. Three potential models will be described including small, medium, and large school models. Wright and Sterry (1983) classified the three programs in the following manner:

- Small Program: These programs often have one or two technology education teachers and less than 160 students in the technology education program at any one time.
- Medium Program: These programs often have two or three technology education teachers and approximately 160 to 300 students enrolled in the technology education program.
- Large Program: These programs often have four or more technology education teachers and over 300 students enrolled in the technology education program at any one time.

The task of designing and implementing a manufacturing program can be difficult. Recently, this task has become considerably easier due to a dramatic increase in the availability of resources such as state curriculum guides, textbooks, professional publications, implementation workshops, and useful media. Still the majority of the responsibility of *what* is to be taught and *when* often rests with a single classroom teacher.

## The High School Manufacturing Sequence

As the teacher(s), entire staff, and/or curriculum specialists begin the task of organizing and implementing a manufacturing program, they should consider an instructional model where students are first introduced to manufacturing technology, then allowed to specialize in related manufacturing systems. Curriculum planners and implementors should also consider an instructional model that progresses rapidly from teacher-directed experiences to student-directed, synthesis courses. In this way, students are encouraged to assume a great role in the teaching/learning process at the higher levels.

Introductory courses can support and encourage the students to develop basic understandings of materials, processes, design, and management systems. Although the introductory course may be primarily teacher-directed, it should be designed to feature numerous open-ended problems related to general manufacturing principles, concepts, and systems.

Specialization courses should allow learners to select courses based on particular student strengths or interests. These courses should be designed to study manufacturing materials and processes, design concepts and practices, and production systems in greater detail. The coursework should also include creative problem solving and small group activities.

Synthesis courses should be designed to allow students to integrate content addressed in earlier courses, life experiences, and other disciplines. This experience should help students apply and build upon prior knowledge related to technology, math, science, business, humanities, psychological and social sciences, and management. As suggested, this experience often takes the form of a manufacturing enterprise where students see manufacturing from a global perspective.

## Small School Programs

In a small school, it may be desirable to limit the manufacturing program to an instructional model similar to the one shown in Figure 5-17. This small school model includes an introductory, specialization, and synthesis course. This gives students the opportunity to participate in an exploratory experience (Introduction to Manufacturing), one specialization course (Manufacturing Materials and Processes), and a capstone course (Manufacturing Enterprise). The synthesis course may also increase the teacher's opportunities to implement new instructional strategies and enriched content.

Because there is no specialization course dedicated to product or system design, the topic of design should be emphasized in the introductory and enterprise courses. Another possibility would be to have the specialization

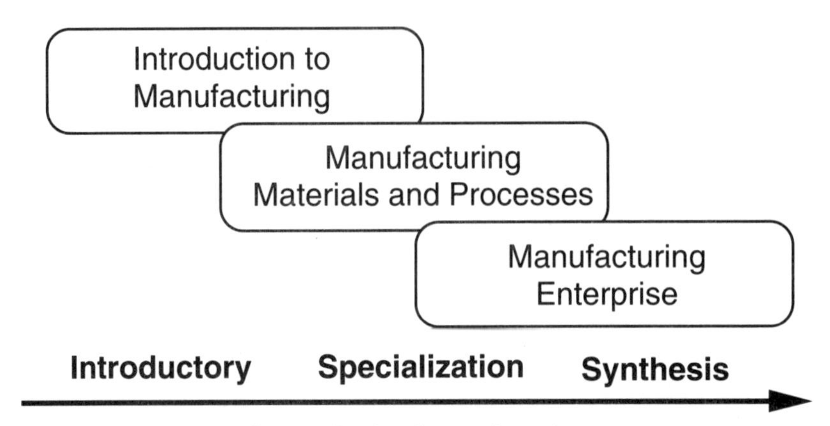


*Figure 5-17: Model of a small school manufacturing sequence.*

courses rotated with the design courses offered during alternating semesters or years. That scheme would allow students to take a materials and processes course one semester and either a product or system design course the following semester.

## Medium and Large School Programs

Typically, bigger schools can offer a wider selection of courses due to more elaborate resources and options. These programs feature more teachers and students, enhanced facilities, more equipment and instructional materials, and flexibility in scheduling. Therefore, students may benefit from a broader range of course offerings. Unfortunately, Seymour (1990) points out that the addition of new specialization or synthesis courses can emphasize problems of overlapping content and laboratory activities. Care must be taken to clearly identify (and separate) course goals, content, and activities.

Teachers in larger programs must also prepare themselves to cover many of the complex or sophisticated topics found in the specialization or synthesis courses. Because material in these courses is often technical in nature a knowledge and understanding of cutting edge or emerging technologies is required. Upper level courses should also include more problem solving, cooperative learning, role playing, and other interactive learning strategies. This creates a need for advanced teaching strategies.

The proposed medium school model (refer to Figure 5-18) is similar to the recommended small school model in that it retains the suggested

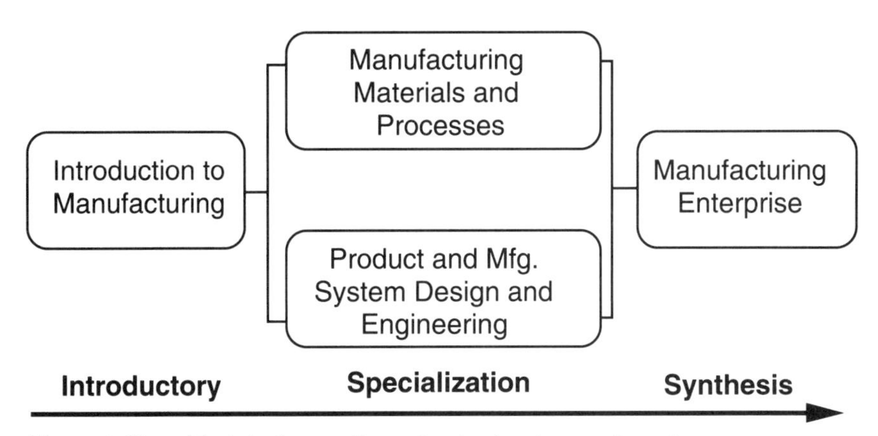


*Figure 5-18: Model of a medium-sized school manufacturing sequence.*

introductory and synthesis courses. However, a second specialization course has been added to the recommended sequence. The additional course expands the student's opportunity to study manufacturing practices used to design, engineer, and specify products and to develop, engineer, and operate manufacturing systems.

The proposed large school model is also similar to the recommended small and medium school models, Figure 5-19. A third specialization course has been added to the recommended sequence by splitting and expanding of the product and manufacturing system design themes into two courses: Product Design and Engineering and Manufacturing System Design and Engineering. Wright and Shackelford (1991a) suggest the separation of the design component into two distinct courses would enhance the program's ability to meet students' needs.

When implementing any new program (especially a program in a large school), scheduling and prerequisites are critical issues. Questions often arise such as "which course(s) should be offered at the freshman and sophomore levels?" or "should this course be a prerequisite for that course?" The combinations of possible manufacturing courses becomes even more complicated when you take into consideration related courses offered in other clusters and disciplines. While certain courses are more appropriate for freshmen versus seniors, improper prerequisites and scheduling have been known to seriously hinder a program's development. In brief, arrange the scheduled courses carefully with minimal restrictions.

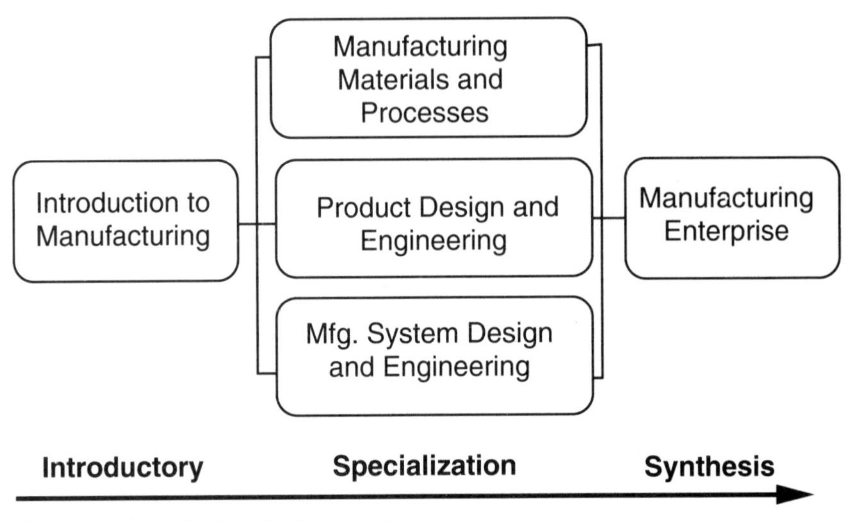


*Figure 5-19: Model of a large school manufacturing sequence.*

# TRENDS AND DIRECTIONS IN MANUFACTURING TECHNOLOGY

It is difficult to address the issue of trends in a yearbook chapter because much of the information is dated before the book is published. Current issues or trends are best addressed in professional journals or conference presentations. However, three important issues will be addressed in this chapter: a) the matter of machine-centered courses (e.g., CAD, robotics, CNC) versus integrated courses; b) the movement towards interdisciplinary learning experiences; and c) curriculum change and the corresponding need for staff in-service and public relations programs. This section will explore these issues and review their implications for the study of manufacturing.

## Machine-Centered vs Integrated Courses

A simple review of state and local curriculum guides and professional journals (articles and advertisements alike) will illustrate the rapid growth in equipment-centered courses. However, in a modern technological society it is not enough that individuals merely be acquainted with a particular machine and its capabilities. The study of work and technology must go beyond a mere device and acquaint students with a) an understanding of the appropriate systems involved and b) how the equipment is used in the system. Individuals must understand the system in which they work in order to be an active participant in the system and society. As Waetjen (1985) notes, "Operating machines may teach us much about technological processes and materials, but it falls short of communicating to us the 'social stuff' of personality." (p. 8)

National studies indicate that schools are not adequately preparing students for a technological society. Students are graduating from our secondary schools reasonably unsophisticated in their understanding of technology and, in particular, manufacturing. More and better educational experiences are required as technology becomes more complex. However, the study of manufacturing must go beyond the study of its machinery. Schools should provide an integrated study that emphasizes the interactions between the people, machines, other bodies of knowledge, and systems of technology. At the high school level, the curriculum should encourage students to assume an active role in the problem-solving, creative thinking, design, materials and processes, production, and management experiences. Thus, a dedicated course in CAD, robotics, or CNC may be just as out-of-place in a contemporary program as Woods IV or Machine Shop 101 was in a contemporary I.A. program.

Technology education is commonly described as that part of the school's general education that teaches students to understand, use, and control technology. To offer a course in CAD (i.e., one that simply replaces the drawing board and T-square with a computerized system) denies the student the opportunity to study how computers fit into the manufacturing and personnel systems of industry. CAD should be used as a resource to support the study of product design and other developmental tasks. Manufacturing courses for the secondary schools should be similar to those described earlier in this chapter. Each of the courses integrate "high technology" topics and systems into the manufacturing curriculum. But, the assumption was that "high tech" equipment supports the curriculum—it isn't the curriculum.

## Interdisciplinary Learning

Interdisciplinary learning is a major challenge facing education in the 1990s and beyond. This movement seeks to address the relationship of math, science, language, humanities, and technology concepts in the curriculum. Several national studies have advanced the concept of integration among all disciplines. The American Association of Science, in their report entitled *Project 2061: Science for all Americans,* states that:

> "The terms and circumstances of human existence can be expected to change radically during the next human lifespan. Science, mathematics, and technology will be the center of that change—causing it, shaping it, responding to it. Therefore, they will be essential to the education of today's children for tomorrow's world." (p. 1)

Changes are being suggested in the math curriculum as well. The National Council of Teachers of Mathematics released two publications entitled *Curriculum and Evaluations Standards for School Mathematics* (1989) and *Professional Standards for Teaching Mathematics* (1991). These documents encourage math teachers to enhance the learning environment in which mathematics is taught by using cooperative learning, problem solving, and relating math to "real life applications" instead of teaching isolated concepts. These suggested changes offer a strong rationale for integrating the study of mathematics and technology concepts. In another familiar report, *A Nation at Risk* (1983), the National Commission on Excellence in Education asserts that the mathematics and science now offered to students would be greatly enriched if teachers were to incorporate technological content.

These studies and reports stress the doing, using, and applying of knowledge. The key themes in technology education involve *doing* words. Technology education integrates learning by doing, creating, inventing, problem solving, planning, experimenting, and producing. Students routinely expand their creativity and problem-solving abilities as they explore technological systems (ITEA, 1990).

Wescott (1991) links the integration of math, science and technology to three important assumptions:

> First, there is a strong linkage between the study of technology education and the appropriate application of math and science concepts. Second, technology education can, with an activity-based methodology, provide an excellent means of integrating math, science and technology into a relevant, meaningful, and functional learning experience. Third, the current emphasis on mathematics and science provides a unique opportunity for technology education to establish itself as a viable discipline to be studied by all students. (p. 2)

These assumptions possess a high degree of relevance to technology education teachers and professionals at all levels.

Students in manufacturing programs apply language, math, science, fine arts, business, social science, technology, and other principles. The technology teacher should actively seek the input, cooperative efforts, and participation of teachers and students from other disciplines. This will lead ultimately to the strengthening of the manufacturing program, enhance the learning of all students involved, and develop a greater appreciation of the technology program in the high school.

## Staff In-Service and Public Relations Programs

As more manufacturing programs are implemented across the country, there is a corresponding need for program transition, revitalization, and staff in-service activities. In-service training is the system used to facilitate change in staff, curriculum, and supporting resources, Figure 5-20. Change may occur in an entire program or simply staff, individual courses, teaching strategies or philosophies, or learning activities.

Seymour (1990b) points out that successful in-service activities are the result of extensive planning and careful organization. One must decide what to cover and how it should be organized to facilitate the objectives of an in-service. A comprehensive in-service program for implementing change includes the following elements:

- an established curriculum,
- competent training staff,
- an identified agenda and time schedule,
-

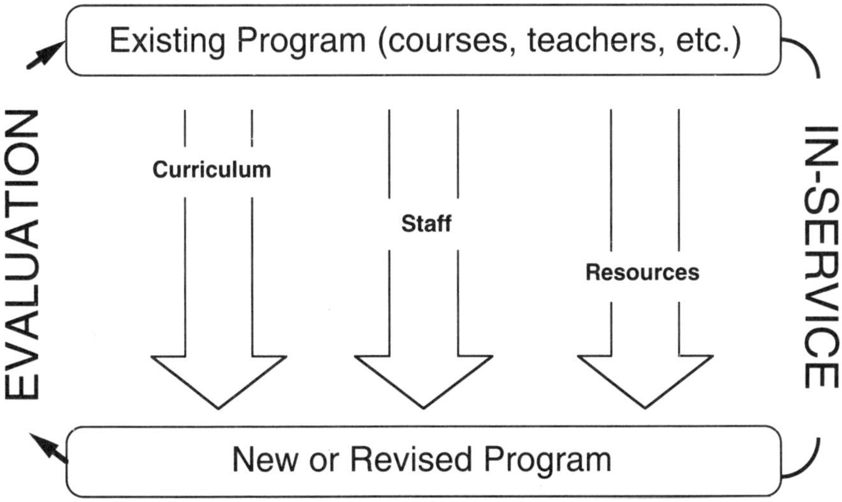

Existing Program (courses, teachers, etc.)
EVALUATION
Curriculum
Staff
Resources
IN-SERVICE
New or Revised Program

*Figure 5-20: Model of an in-service agenda (adapted from Shackelford and Seymour, 1987).*

- training facilities,
- required support resources and materials,
- dissemination of results,
- a program evaluation system,
- program follow-up, and
- in-service participants with known characteristics.

Acknowledging these elements, the in-service director and staff must define the purpose and scope of the in-service, prepare workshop materials, select and prepare trainers, schedule the actual training, focus on facility and resource demands, etc.

In secondary schools, in-service activities are required to help the staff and community develop and/or implement a new or revised curriculum. The school community involves other teachers, administrators, students, parents, and local business and industry. Implementing "change" is a group effort requiring input and involvement from the entire school community. Technology teachers must be encouraged to work cooperatively with fellow staff members and other teachers in the school. They must be encouraged to form teaching teams, not cliques, and to utilize individual and group strengths. In addition, they should also give up our traditional attitudes that this is "my lab" or "my course" for the perception that this is "our new facility" or this is a new course "we offer."

A cooperative effort should extend beyond the technology program. Teachers should work closely with the school administration and counselors to communicate the goals and objectives of the manufacturing program. Formal communication needs to include explanations and examples of what students will be doing in the new manufacturing courses. Effective public relations improve the ability of counselors to explain the advantages of the course to students and parents. It may also enhance prospects for acquiring necessary resources from the administration.

The communication process should be formalized into an internal (in school) and external (local community) public relations program. The program might publicize programs objectives plus interdisciplinary linkages, student's accomplishments, on-going activities, and involvement in local service activities. For many technology teachers, the local community contains numerous untapped resources. Developing an effective internal and external public relations program may help expand the manufacturing program beyond your wildest dreams.

# SUMMARY

Manufacturing technology is a primary technological system. The study of manufacturing technology can provide a foundation for technological literacy and support the development of insights into how technological concepts, processes, and systems can be used to solve practical problems. At the high school level, this study can a) help provide students knowledge and skills necessary to understand, use, and control manufacturing systems; b) address manufacturing's role in modern society and interrelationships with other areas; and c) promote an understanding of the consequences of manufacturing on society, economies, and the environment.

In the manufacturing sequence, emphasis is placed on the "doing words" such as creating, planning, problem-solving, experimenting, and producing. Student experiences involved the safe use of tools; application of technical concepts; processing materials, energy, and information; designing products and production systems; and testing management concepts and procedures. In addition students benefit from opportunities to uncover and develop individual talents; sharpen critical thinking, problem-solving, and decision making skills; and function as individuals and team members in cooperative activities.

Several models for studying manufacturing technology were introduced and described. The high school curriculum was operationalized through a planned series of courses designed to promote the study of manufacturing technology and meet the needs of the students, society, and the body of knowledge to be studied. Courses described in detail included: Manufacturing Materials and Processes, Product Design and Engineering, Manufacturing System Design and Engineering, and Manufacturing Enterprise. The description of each identified course included selected content, modules, learning experiences, instructional sequences, and interdisciplinary activities.

Recommended program structures for implementing the study of manufacturing technology in different school settings (e.g., small, medium, and large schools) were also discussed. These suggestions were followed by a review of trends and issues in manufacturing technology. One trend involved the enhancement of the learning environment through the implementation of interdisciplinary learning. Two issues included the study of manufacturing through machine-centered courses as compared to integrated content courses and the importance in-service and public relations programs in the implementation of a new manufacturing technology program.

# *REFERENCES*

Ball, E. (1989). Technology Programs -- Ann Arbor, Michigan. In *Technology education in action—outstanding programs*. Reston, VA: International Technology Education Association.

Beranek, S. (1984). The student based enterprise: A success story. *Manufacturing Forum, 10* (3), 3-8.

*Curriculum and evaluation standards for school mathematics*. (1989). Reston, VA: National Council of Teachers of Mathematics.

Daiber, R., & Erekson, T. (1991). *Manufacturing technology today and tomorrow*. Mission Hills: Glencoe/McGraw-Hill.

Deming, W. (1991, November 6). Business in the global market place. *Chicago Tribune*, p. 25.

Durkin, J. (1984). *Technology lab 2000*. San Diego, CA: Transtech Systems—Division of Creative Learning Systems.

Fales, J., Sheets, E., Mervich, G., & Dinan, D. (1986). *Manufacturing a basic text*. Encino, CA: Glencoe.

*Industrial technology orientation curriculum guide*. (1989). Springfield, IL: Illinois State Board of Education.

*Introduction to manufacturing technology*. (1989). Indiana Industrial Technology Education Curriculum Guide, Indianapolis, IN: Indiana Department of Education.

Komacek, S., Lawson, A., & Horton, A. (1990). *Manufacturing technology*. Albany, NY: Delmar.

Kreitner, R. (1983). *Management*. Boston: Houghton Mifflin Co.

*Manufacturing enterprise*. (1989). Indiana Industrial Technology Education Curriculum Guide, Indianapolis, IN: Indiana Department of Education.

*Manufacturing materials and processes*. (1989). Indiana Industrial Technology Education Curriculum Guide, Indianapolis, IN: Indiana Department of Education.

*Nation at risk*. (1983). United States Department of Education, Washington, D.C.: National Commission on Excellence in Education.

*Product and manufacturing system design*. (1987). Indiana Industrial Technology Education Curriculum Guide, Indianapolis, IN: Indiana Department of Education.

*Professional standards for teaching mathematics*. (1991). Reston, VA: National Council of Teachers of Mathematics.

Rye, E. (1989). Technology Programs—Bellevue, Washington. In *Technology education in action—Outstanding programs*. Reston, VA: International Technology Education Association.

Seymour, R. D. (1990a). Conceptual models for communication in technology education programs at the high school level. In J. A. Liedtke (Ed.), *Communication in Technology Education* (pp. 62-91). 39th Yearbook, Council on Technology Teacher Education, Peoria, IL: Glencoe.

Seymour, R. D. (1990b). Workshop director's handbook: A guide for conducting in-service workshops for secondary school manufacturing courses. Muncie, IN: Center for Implementing Technology Education, Ball State University.

Shackelford, R., & Seymour, R. (1987, March). *Conducting technology education in-service training programs*. Paper presented at the meeting of the International Technology Education Association, Tulsa, OK.

Shackelford, R. & Todd, R. (1979). A flexible equipment system to support innovative programs in technology. *Man/Society/Technology*, *39* (7), 20-23.

Shackelford, R. L. & Wescott, J. (1991). *Manufacturing materials and processes*. Muncie, IN: Center for Implementing Technology Education, Ball State University.

Shackelford, R., Wright, T., & Haynes, F. (1987). *Industrial technology education—A guide for facility planning*. Indianapolis, IN: Indiana Department of Education.

Squier, T. (1984). Managing your student enterprise. *Manufacturing Forum*, *8* (3), 3-6.

*Technology education its opportunities*. Reston, VA: International Technology Education Association.

*Technology education . . . the new basic*. (1988). Reston, VA: International Technology Education Association.

Tuma, R. & Fisher, M. (1991, November, 6). *Workforce development. Chicago Tribune*, p 27.

Waetjen, W. (1985). People and culture in our technological society. In *Technology education : A perspective on implementation*. Reston, VA: International Technology Education Association.

Wescott, J. (1991). Integrating mathematics/science/technology: A challenge for educators. *Indiana Technology Education Newsletter*, Muncie, IN: Indiana Industrial Technology Education Association. *4* (1), 1-3.

Wright, R. T. (1990). *Manufacturing systems*. South Holland: Goodheart-Willcox.

Wright, R. T. & Shackelford, R. L. (1990a). *Manufacturing system design and engineering*. Muncie, IN: Center for Implementing Technology Education, Ball State University.

Wright, R. T., & Shackelford, R. L. (1990b). *Product design and engineering*. Muncie, IN: Center for Implementing Technology Education, Ball State University.

Wright, R. T., & Sterry, L. (1983). *Industry and technology education: A guide for curriculum designers, implementors, and teachers*. San Marcos, TX: Technical Foundation of America.

CHAPTER 6

# Manufacturing Technology In Teacher Education Programs

Dr. James E. LaPorte (Associate Professor)
*Virginia Tech, Blacksburg, VA*

"We teach as we have been taught" is an old adage in many educational circles. Yet, the adage can be backed by empirical evidence. A program designed to prepare high quality manufacturing teachers must involve an exemplary model or "stencil" that is transferrable to the public schools. It should not be an assemblage of disjointed courses and fragmented experiences based on the assumption that the teacher will be able to synthesize them and somehow put them into teachable terms for future students. At the same time, the manufacturing experience at the collegiate level must be challenging to the mature, aspiring professional teacher. This is the dilemma that faces teacher educators responsible for the manufacturing sequence in their programs and departments.

## The Diversity of Teacher Education Programs

Diversity characterizes the manner in which technology education teachers are prepared. There are three significant reasons for this variation. First, there are the varying philosophies that undergird what the prospective teacher should experience as he or she moves through the program. Quite often, the differences in philosophy and resultant practice have been augmented by the transition from industrial arts to technology education.

Second, a number of initiatives during the 1980s called for the reform of teacher education. Notable among them were the reports of the Holmes Group (1986) and the Carnegie Task Force (1986). Both initiatives are similar in that they propose a teacher preparation approach modeled after professional schools like medicine, dentistry, and the legal profession. Using this model, the aspiring teacher completes an undergraduate program in the discipline of choice and then enters a professional school which provides an in-depth study in educational theory and practice.

The Holmes Group was specifically targeted toward research-oriented institutions. Invitations to participate were offered only to those institutions that met prescribed criteria for research productivity. The result has been that institutions subscribing to the Holmes Group are, by and large, preparing teachers differently from those who are not embracing the Holmes Group ideals. Since technology education is not generally considered to be a "discipline" in the academic sense of the word, some institutions have had to resort to finding a discipline within the institution that most closely matches their program of study and abandon their original program.

As a corollary to the Holmes Group approach, new programs are being developed to articulate with community colleges and other two-year institutions. Students choosing this route would complete a planned program at the two-year institution, then continue their academic work at the four-year institution. All credits earned during the two-year experience would count toward a degree at the four-year institution. Again, this adds yet another dimension of diversity to the preparation of manufacturing teachers.

The community college portion of the program would most certainly place the prospective technology teacher in classes in which the majority of students are bound for careers in industry. But unlike technology education programs in four-year institutions, where the emphasis is on non-teaching careers, such students would not even have the advantage of interacting and receiving guidance from technology teacher educators. Miller (1991) presented models for community college articulation programs and several alternatives for the preparation of technology teachers.

A third factor adding to the diversity in the preparation of technology education teachers is the mixture of students within the program itself. There are few "pure" technology education programs that exist today; in fact there are fewer than ten (LaPorte, 1988). Most programs prepare future technology teachers along with students who do not aspire to teach. As a result, courses are often shared with similar majors. That is, prospective teachers take many of the same courses as do their non-teaching counterparts. On the one hand, this can be an advantage since courses aimed primarily for industry-bound graduates are often well-supported by industry and may therefore be more current in terms of content and equipment. On the other hand, such mixed courses provide little if any opportunities for references to pedagogy. Further, the appropriate manner in which to structure curriculum for students headed for careers in industry is usually quite different from what may be appropriate for prospective teachers. In addition, since the industry-bound students are often the majority, more attention may be paid to meeting their needs.

These three reasons plus factors such as geographic region, size of the program, resource availability, and the nature of the institution, the challenge of preparing technology education teachers in general, and manufacturing teachers specifically, is indeed great. However, the recent standards for the accreditation of technology teacher education programs established by CTTE and implemented in concert with the National Council for the Accreditation of Teacher Education may lead to a measure of consistency across programs (Starkweather, 1991). Nonetheless, the diverse nature of technology teacher education must be considered in proposing program models.

### Teacher Education and the Public Schools Compared

The purpose of manufacturing education in the public schools is to increase the literacy of the learner about our global manufacturing system. The purpose of manufacturing programs in teacher education is to increase literacy about manufacturing *and* prepare a competent teacher of manufacturing. One of the biggest challenges at the teacher education level is to provide meaningful learning experiences to the aspiring teacher that match his or her level of intellectual maturity while at the same time providing a model of manufacturing technology that is transferable to the context of the public school.

This challenge could be more easily met if the prospective teacher entered the collegiate teacher education program with a reasonably sound educational experience in manufacturing obtained at the secondary level. However, this is simply not the case. Few technology teacher education majors have had prior coursework in technology education upon entering a teacher education program and only a small proportion of those have had a quality experience in manufacturing. Though this situation may improve in years to come, a truly articulated program would result only if technology education, specifically manufacturing, becomes a required experience at the secondary level for everyone.

Thus, the coursework that the prospective manufacturing teacher takes should share a common structure with the program delivered at the lower levels. However, the learning experiences must match the maturity level of collegiate learners and be academically challenging.

## THREE COMPONENTS OF A TEACHER EDUCATION PROGRAM

Years ago, Nelson (1962) proposed that there were three components of an industrial arts teacher education program: professional education, general education, and technical education. These three categories seem appropriate today with the exception of the last. Contemporary technology education programs are much more inclusive than the term technical denotes.

"Technical" is an important part of technology education and is included in "technology," but it is a delimited subset of the latter. At the same time, a distinction should be made between a purely academic study of technology as compared to one that involves actual technological practice (such as technology education). Therefore, the term "technical/technological" has been used herein.

The *technical/technological component* includes those courses that specifically deal with the discipline of technology. In simple terms, the technical sub-component includes the learning experiences that focus on the *how-to* and encompasses the use of tools and machines, designing and developing technological systems, and processing materials. The technological sub-component includes the learning experiences that deal with the *about* of technology. It encompasses such elements as systems analysis, conceptualizing problems, and considering the consequences of technology. As one might suspect, there is no absolute demarcation between the technical and technological component and it is thus represented by dashed lines in Figure 6-1.

The *professional component* includes coursework which encompasses the theory and practice of teaching. Included in the professional component are methods courses, field experiences, and student teaching. In addition, the professional component includes the pedagogy and teaching experiences that occur in the technical/technological category. For example, a professor teaching a technical/technological course may refer to the transferability to the public school setting in the delivery of demonstrations, the organization of a course, or the management of a particular laboratory situation.

The *general component* involves any courses that are designed to provide broad, essential, educational experiences and learnings that are considered to be useful to everyone regardless of their career aspirations. These courses provide some of the foundational elements for the study of technology. This component includes the "core" requirements of the institution (i.e., social studies, humanities, language, mathematics, and the sciences) plus the requirements from elective areas.

A generalized mix among the three components of a conventional teacher education program are depicted in the upper portion of Figure 6-1. The illustration is adopted from the earlier work of Nelson (1962). In this conventional model, students would complete the required coursework within four years. Attention to the core requirements is concentrated at the beginning of the program and the technology specific coursework is con-

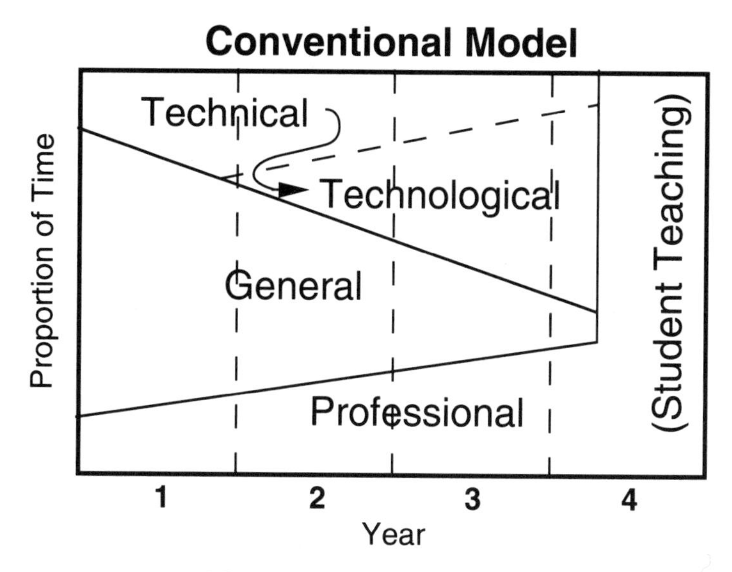


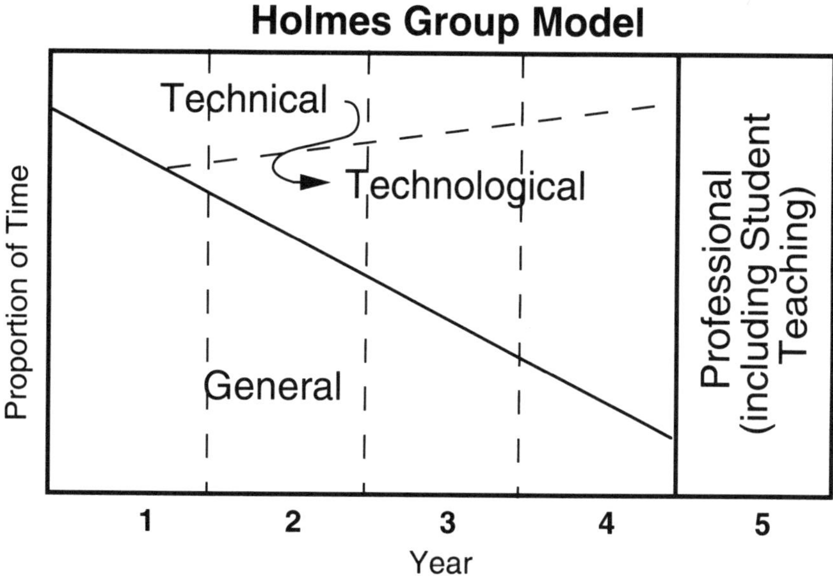


*Figure 6-1: Comparison of time spent in the three components of teacher education by year—Conventional versus Holmes models.*

centrated at the end. Typically, the final semester is devoted to student teaching. Graduates of such a program would be awarded a Bachelor's degree.

The lower portion of Figure 6-1 depicts a possible model integrating the three components within a Holmes Group-type program. During the first four years, study would be focused on the "discipline" (technology) and on the core requirements of the institution. The fifth year would be devoted exclusively to professional study in education, field experiences, and student teaching. Graduates of the program would be awarded a Bachelor's degree in a technology-based discipline and a Master's degree in education (or at least make substantial progress towards such a degree).

Keeping in mind that Figure 6-1 shows only one possible implementation of the Holmes Group ideals, there are some advantages as well as disadvantages to such a program. Since all pedagogical (professional) as well as the student teaching experience would occur during the fifth year, more time is available for study in technology (and potentially, manufacturing technologies). On the other hand, what might constitute the discipline of technology is subject to interpretation. Some institutions might merely preserve the elements of their former industrial arts undergraduate program and expand the content to fill the extra time. Yet others might decide that engineering technology, industrial technology, or even engineering might be the appropriate discipline for future teachers of technology. There is some evidence of movement in this latter direction by a few institutions.

New models for preparing teachers in general and technology teachers specifically will no doubt increase the already significant diversity that exists. As far as the Holmes Group goes, however, its impact has been primarily on research-oriented, largely land grant, institutions. But this in itself could further confuse the manner in which technology teachers are prepared, adding yet another dimension of diversity.

## The "Hour Glass" Model

In an earlier era, teacher education students often commenced their program of study with specific, material-based courses such as woodworking, metalworking, plastics, etc. Unavoidably, the implicit context of such an approach was that of the "basement workshop" or a cottage industry. Efforts to create a different, more contemporary context through more teacher-centered (mostly verbal) experiences typically failed. Today, such material-based courses have been replaced with materials and processes courses, providing an *intra*-disciplinary experience. In such courses, the learner is provided with the opportunity to learn processes conceptually and to better comprehend the commonality of processes across a variety of materials. Conceptual approaches to teaching technology have gained widespread acceptance within education and technology education specifically, at least since the introduction of the Industrial Arts Curriculum Project (Towers, Lux, and Ray, 1966). Conceptual learning (as opposed to factual learning) offers several advantages. These include bringing meaning to learning, enhancing student coherence to information and experiences, and enabling lifelong learning (Loucks-Horsley, 1990). Nonetheless, it can be a giant conceptual leap between the materials/processes experience and the development of a complete manufacturing system.

Establishing a viable context for manufacturing experiences is important, perhaps critical. This context should not be provided through verbalization by the teacher, but by practical, experiential learning. To illustrate this point, consider a person who has never seen an automobile before and wishes to learn about it. One approach would be to start with the carburetor and study its functions. Then the learner would study the ignition system. Each subsystem of the automobile would ultimately be studied in a planned order. Finally, the entirety of the automobile would be presented and the interrelationships of all the subsystems shown. But during this example, the learner has had no idea of the purpose and function of each of the subsystems relative to the automobile itself. Instead, these subsystems were learned in isolation and without a context.

A more defensible approach would be to show the automobile in total, thereby establishing the context for the learning that is to follow. The central mission and function of this common transportation device (i.e., the auto) would be covered. In time, each of the subsystems would be studied within this known context. Once this had been accomplished, the automobile in its entirety would be revisited so the learner could more readily understand how each of the subsystems work in relationship to one another. Using such an approach, instruction starts with the general, moves to the specific, and then concludes with the general. This general-specific-general approach to instructional sequencing is depicted in Figure 6-2 and is referred to as the "Hour Glass" model.

This model lays the basis for the contemporary preparation of a manufacturing educator. The students should be provided a general overview of the production sector—its importance in society, function, day-to-day operations, etc. This study would be subdivided into key themes with appropriate coursework structured to address the specific topics. Typically themes include materials and processing, design, manufacturing enterprises, and modern manufacturing technologies.

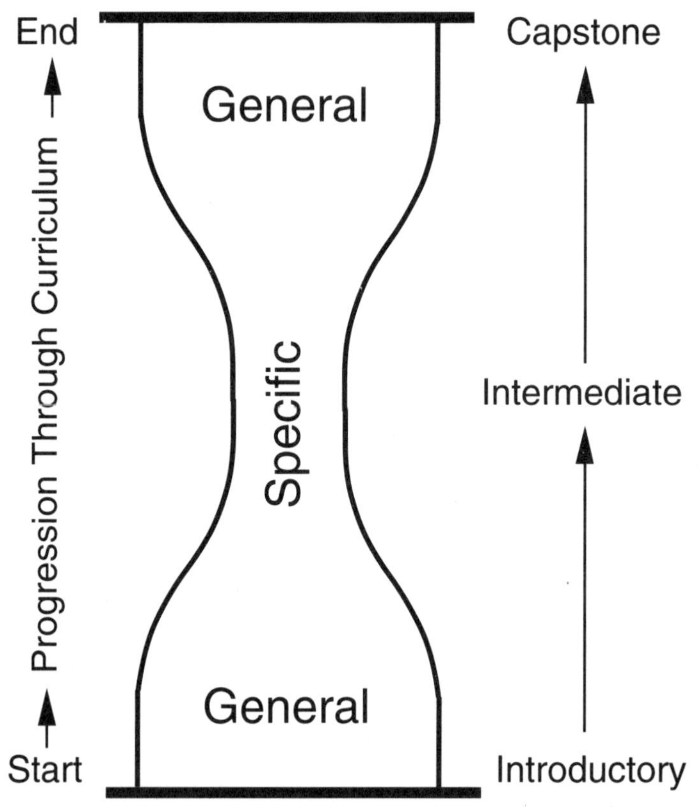


*Figure 6-2: An "hour glass" model of sequencing a curriculum.*

# MANUFACTURING EDUCATION AND THE TECHNICAL/TECHNOLOGICAL SEQUENCE

Today's collegiate student in technology education will likely take 10-18 semester hours in courses related to manufacturing, production, and/or material processing themes. The initial experiences should provide an introductory view of manufacturing. As mentioned, many collegiate students have little time to take elective classes (technology education, etc.) at the secondary level so may have a limited background in manufacturing technology. This section outlines several courses for the technical/ technological component of a manufacturing sequence and suggests content, activities, and important student experiences.

## The Introductory Experience

Adhering to the Hour Glass model, manufacturing technology would be introduced by providing the students with an actual, hands-on manufacturing experience. One way to accomplish this would be for the instructor to establish the requirements for an entire line production activity and have novice students assume responsibility for each workstation. The experience would be brief and could be included in courses taught early in the student's program of study (i.e., classes like "Introduction to Technology," "Introduction to Production," or as the initial activity in a "Materials and Processes" course). Though the experience is short in duration, it results in the learner having a conceptual model of what manufacturing is and involves. Further, the experience can thereby introduce the various processes and systems and related experiences which will follow. If successful, it establishes the context for the upcoming experiences in design, research and development, problem solving, manufacturing enterprises, etc.

## Materials and Processes Experience

Once the conceptual model for manufacturing has been established through a structured introductory experience, the learners should move into the second stage of the Hour Glass model, learning the "specifics" related to manufacturing. A principal segment of any teacher preparation program includes coursework that features the nature and properties of materials and how they are processed.

Just as it is important to establish the proper context for manufacturing, it is likewise imperative to set the proper context for materials and processes experiences. Unfortunately, some materials/processes courses are taught without a context. For example, a common approach to teach the properties of materials is to require students to perform laboratory tests to determine such properties as tensile strength, compressive strength, hardness, and so forth. Students perform the required exercises in absence of a frame of reference into which the understandings can be fit or from which meaning can be obtained. A defensible context for the materials and processes experience is the manufactured product. Instead of having students perform routine tests, problems can be readily developed so that the students are challenged to determine which material is best suited for a particular application in a product. Then, the testing and analysis is done to solve the problem. Thereby, the purpose for material testing and analysis takes on real meaning.

In learning about the processes of manufacturing, some skill in the use of machines and materials is a natural and desirable outcome. However, conceptual understanding rather than skill should be the primary goal. Over the years, the technology education profession has reached a degree of consensus on what these concepts should be. The Industrial Arts Curriculum Project (Towers, Lux, & Ray, 1966) rationalized that combining, forming, and separating were totally inclusive, mutually exclusive, and operationally adequate in describing how materials are processed. These concepts have met the test of time to a large extent. However, these three concepts did not include processes that change the internal structure of materials such as hardening, tempering, and annealing. Therefore, conditioning was added as a fourth concept. Wright (1990) added assembling, casting and molding, and finishing, bringing the total to seven categories. Various authors have since identified different organizers and structures. In practice, the differences in how many higher-order concepts are studied are rather moot sincc it is principally a matter of perspective and has little real impact on how the course is delivered.

The important thing about a conceptual structure for the processing of materials is that it provides a model for the prospective teacher to organize and deliver instruction. New processes can be readily integrated into the over-all model as they are developed. In addition, outmoded practices can be removed without altering the basic model.

Such a structure also allows for the commonalties among processes to be called to the attention of the learner. For instance, learners should clearly understand the similarities among the extrusion of clay, plastics, and metals. In another example, they should recognize the similarity between slush casting of vinyl plastisols and the slip casting of ceramic products.

The outcome of the materials and processes experience should be the development of a conceptual database of materials and processes in order that, when confronted with a problem like producing a hole in a piece of stock, students don't automatically think of using a twist drill and drill press. Instead, they draw upon their prior experiences and knowledge of material properties and processes for producing the hole using the most appropriate means. Typical examples include punching, electric discharge machining, boring, induced fracture, laser burning, etc. Their experiences also help provide a reference for innovation and inventiveness. Unfortunately, once established, this conceptual approach to materials and processes is quickly lost if the laboratory experiences are structured in such a manner that using a twist drill in a drill press is always the only solution to producing a simple hole.

A manufactured product serves as the first-order context for materials and processes. But, there are second-order contexts that play an important role in structuring instructional experiences in materials and processes. A possible context-based model for the materials and processes experience is illustrated in Figure 6-3.

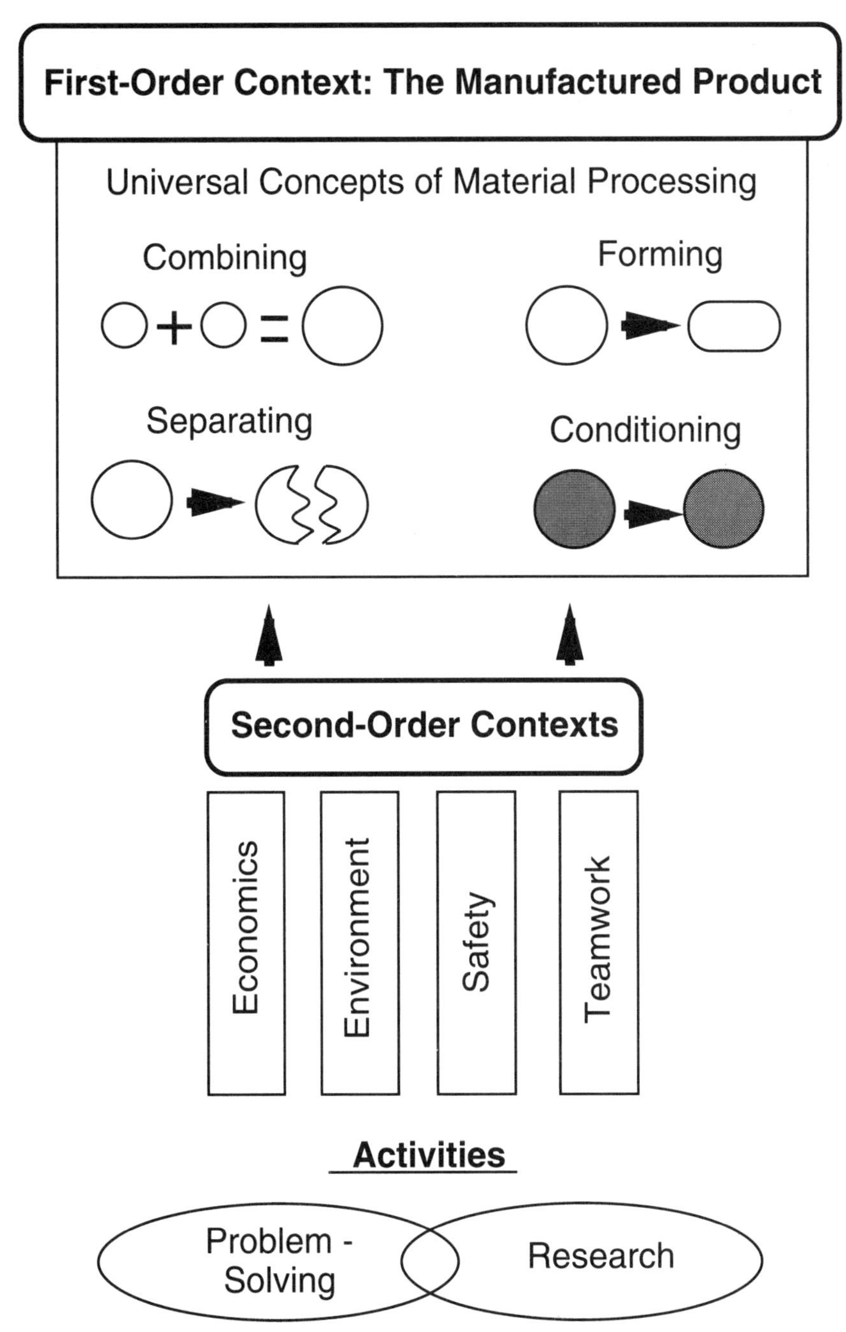


*Figure 6-3: A contextual model for studying materials and processes.*

One second-order context is economics. Over the course of history, economics has played a significant role in technological development. It follows that economics should be included in the materials and processes experience as well. The decision to use a particular process and material must consider the costs involved. These costs go beyond the expense of the machines and equipment, but also include the amount of skill and associated labor costs. The same is true concerning the selection of materials. Clearly, the inclusion of automation and its cost effectiveness is another topic worthy of consideration.

A simple economic-based learning experience might focus upon a pneumatic nail driver. The problem would involve determining how many nails are required to be driven before it is economically feasible to use the nail driver instead of a standard hammer. Getting the tool ready to use, the cost of the nails, and the depreciation on the machine are several of the factors that could be considered. Not only are such learning experiences sound, general education experiences, but they readily transfer to the decisions the prospective teachers will have to make both in the management of a laboratory as well as his or her personal life.

Another second-order context for materials and processes is the environment. Studying how raw materials are changed into standard stock should lead naturally into a study of the sources of raw materials, what pollutants are produced in the conversion processes, whether or not the materials are renewable, how much energy is used in conversion, and the proper disposal of waste and scrap.

But environmental knowledge must not be exclusively cognitive. Technology education laboratories offer excellent opportunities to actually apply sound, environmentally considerate practices—they provide a microcosm of the larger world. Students can apply proper waste disposal and recycling practices to the actual materials they use in the lab. For instance, they should learn to contemplate the consequences of their actions before introducing potentially hazardous materials into the sanitary sewer system. They should also consider what materials will have the least negative impact on the environment as they select among alternative materials.

Requiring students to develop an environmental plan for the manufacturing laboratory is one way in which this concept could be addressed. The result is not only a cognitive understanding, but it reaches into the psychomotor and affective domains as well. Such experiences develop an environmental conscientiousness that might positively influence the public school students with whom future teachers will work.

Safety is yet another second-order context for materials and processes. Safety must be taught with respect to all three domains of knowledge. In the cognitive domain, students learn about the operation of machinery and their limits. In the psychomotor domain, students learn how to actually use machines and equipment safely. In the affective domain, students learn to value safe practices. However, concepts of safe practice must not end when instruction ceases—it must serve as a continual context for what occurs in the classroom and laboratory.

Teamwork is a fourth, second-order context for materials and processes. In modern society, people rarely work autonomously as they did in the past. Modern management practices rely extensively upon teams to reach decisions and solve problems, capturing the notion of synergism (i.e., that the whole is greater than the sum of the parts). The emphasis on team efforts was detailed extensively when announcing the grand opening of the Chrysler Center for Technology.

A manufacturing materials and processes course provides an excellent environment for fostering the team concept. The team concept helps students develop strategies for working together to reach consensus and solve problems, and become interdependent on each other for learning. This context is not taught as an individual unit, but becomes a significant theme throughout the course.

The four second-order contexts of economics, the environment, safety, and teamwork are not totally inclusive of all possible contexts that could be applied in the manufacturing materials and processes sequence. But, like technology itself, they are dynamic. Perhaps more importantly they help separate the good from the excellent programs; the enhanced materials and processes experiences offer the students more than merely facts and data.

Student learning activities in materials and processes coursework should promote both problem-solving and research skills. Problem solving has been a central theme of technology education programs. A key element of virtually all models of the problem-solving process involves research. Research and problem solving should serve as the primary learning activities for all materials and processes coursework, Figure 6-3.

As a result of their study and experiences, students may develop a natural curiosity about materials and processes. Courses should be structured so that students can spontaneously delve into technical research and experimentation. Research experiments can serve as a culminating experience for students in materials and processes. More formally, students should learn about the concepts of experimental control, sampling theory, validity, and reliability through applications such as a mini-research study. As a routine class assignment, students would prepare a proposal and present it to others in the class for critique and suggestions. As part of the requirement, the student would have to choose or develop appropriate apparatus to actually perform the necessary tests and experiments. Then they would proceed to collect data, analyze it, and draw conclusions. Finally, a written report and presentation would be prepared.

Several examples of mini-research questions which have proven to be effective in previous classes include:

1. Which automotive finish is most resistant to gravel damage?
2. Is a glue joint stronger if glue is applied to both surfaces of the joint?
3. Which commonly available mechanical fastener provides the greatest strength in joining particleboard?
4. Which adhesive provides the strongest bond between metals in which there is a constant vibration?
5. What is the sound absorption of selected wood species?
6. To what extent does mildly acidic water absorb lead from soldered plumbing joints?

Excellent references for integrating research problems have been prepared by Maley (1988a, 1988b). Though out-of-print, a book by Earl (1960) had some excellent practical ideas for conducting research on the properties of materials. In addition, the 1987 CTTE Yearbook titled *Conducting Technical Research* (Israel and Wright, 1987) is a valuable reference on the subject.

## The Design Experience

Design is another important theme in modern manufacturing technology. Both products and systems must be carefully planned to match societal, economic, and production criteria. Therefore, a major element in the collegiate manufacturing sequence should address design and related topics (research and development, innovation, etc.).

A contemporary manufacturing program might feature one or more dedicated courses with a design theme. For instance, a course in product design would include student activities that cover sketching, engineering drawings, model making, and prototyping. Assignments would challenge students to create new and improved designs of consumer and industrial products. Another typical course might focus on developing production systems. The course would include the topics related to the design of manufacturing systems—plant layout, tooling design, material handling, preparing production paperwork, etc.

## The Manufacturing Enterprise Experience

The manufacturing enterprise experience is often the capstone course at the collegiate level. It brings together a student's understanding of manufacturing technology with knowledge from his/her general education courses. The prospective manufacturing teacher should leave the enterprise experience with a clear understanding of the managed systems and implementation strategies that are transferrable to a public school situation.

In the enterprise approach, student-managed companies embody the definition of an enterprise as they assume a risk in hopes of making a profit. The students are challenged with designing and developing a product, designing and implementing a production system, and then producing the product within a fixed amount of time.

The time element is a critical aspect of the experience. The class should have a firm, prescribed time for the production of the products. For example, a single two-hour laboratory period at the end of the term could be scheduled. This gives the students a sense of importance that may not be achieved in any other way. The reality of what can happen if "one weak link in the chain" fails is faced by the students on a daily basis. Adding the time element sets a rigid standard for the experience that is analogous to "playing in the game" rather than "just playing catch." What's more, the organizational experience that the students receive is transferrable to their personal life as well as to virtually every occupation. Finally, it should portray a model that is transferrable to the public schools. Too often, manufacturing in the public schools is little more than custom production rather than a true manufacturing experience.

A second key aspect is requiring a true, mass production system. That is, on the designated production day, raw materials flow in one end of the production system while finished products flow out the back of the system. Some component parts or subassemblies of the product may be produced ahead of time, simulating the common practice of subcontracting with outside vendors. In fact, such outside contracting is a perfect opportunity for the students to engage in a "global manufacturing" activity whereby parts for the chosen product are produced by students at a distant school (Novotny, Thomasser, & Chapin, 1990). But the majority of the operations should be done within the confines of the scheduled production session.

A third key aspect is putting the responsibility for success in the hands of the students. To do this, teaming is often used. The class is divided into small groups of five to eight students on each team (or "company"). The students are responsible for all of the decision making and the teacher intervenes only when absolutely necessary. The companies compete with each other, simulating the competition that goes on in the world outside of the university. Like competition in the real world, they become protective of their innovative ideas. The competitive environment also enhances learning because students often go beyond the minimum requirements of the class.

The competitive spirit is equally balanced by a need for cooperation. As the production system is developed, the two or more companies within the class must share the resources of the laboratory with each other. This includes the tools and equipment, production space, and computers, as well as the counsel of the teacher. Again, cooperation comes into play on the production day for a particular company. With teams of five to eight students, there is simply not enough human resources to complete all the operations in the developed system. The companies may be required to "hire" talent from competing companies and assure that they will maintain an acceptable level of quality during the actual production.

In using this small group approach, it is advisable to randomly assign students to a company. This often improves the mixture of abilities, performance levels, and personalities. No doubt, there will be some conflict among the members of the company, but this is one of the advantages of the enterprise approach in that situations often parallel the conflicts that occur in everyday life. Once the students have been assigned to a company, they can determine individual responsibilities plus an organizational structure and job descriptions. This requires the students to become thoroughly familiar with production requirements and what it takes to develop a total manufacturing system.

An enterprise starts with a creative suggestion that is brought to fruition during a product ideation process. Each student formally presents what he or she thinks is a creative idea for a marketable product (that fits announced restrictions such as being producible within the time constraints and with the equipment available). The developed product idea should eventually be presented with an appearance mock-up, design sketches, and similar materials. The members of the company then reach consensus on one or more product ideas and begin the production of the corresponding prototype(s). Through group consensus, a final decision is made on the single product to be produced by the team.

Market research can be integrated in several ways. Data can be collected at the mock-up stage of product development, thereby comparing alternative designs and features. Or, marketing data could be collected on several versions of the same product. In either case, the data should be collected from an identified sample of potential consumers of the product. Attention should be paid to reliability and validity of the data. To collect data, a table could be set up within or outside of a prominent building on the campus.

Students should be responsible for raising their own capital. This forces the group to prepare a budget and do some financial forecasting. It also clarifies the concepts of direct and indirect costs, profit margins, establishing the selling price, and determining the break-even point. One method to raise capital is to sell stock to other students or to the general public. Just as in the real world, some of the profits of the company are distributed to the shareholders upon dissolution of the company.

With the company organized, a product chosen, marketing research completed, and capital raised, the students can proceed to develop the actual production system. This includes preparing working drawings, production plans, tooling, and the development of a quality assurance system. Student efforts during this phase of the course would build on experiences from prior coursework on design and product development. The production of tooling is technically challenging and will likely take the most time. But remember, educational success is related directly to the quality of the experiences that the students have in completing these assignments.

The design of product packaging can begin as soon as the product is finalized. Ideally, each product would be packaged. This involves the development of a production subsystem that parallels most of the steps involved in the system designed for the product itself. As an alternative, the package could be contracted out to members of a communication or graphics course, providing yet another novel experience for the enterprise students.

A trial run of the developed production system is absolutely essential. This step involves the production of a single product with the members of the company walking through the various operations and critiquing their efficiency, accuracy, and safety. A time study of each operation should be incorporated into the trial run.

During the actual production run, students from the competing companies may be hired to assist with the fabrication and assembly efforts. In addition, students and faculty members from within the department and across campus can be involved. These "outsiders" must be trained for their jobs and taught applicable safety practices as well. Not only will this result in sufficient human resources, but it also results in good publicity for the class and the program in general.

After the products have been successfully produced and marketed, the company must be dissolved. All profits must be distributed to the stockholders and the remaining assets of the company must be disposed. One approach to the disposal of assets is to hold an auction, inviting potential buyers of the leftover inventory (screws, finish, surplus materials, etc.).

Throughout each part of the enterprise experience, the pre-service teachers should discuss how their experiences may need to be modified when conducted in a public school. This helps place their experience in a pedagogical context. Further, individual and group assignments along these lines will help reinforce thinking about manufacturing at the lower levels.

Topics for lecture and discussion that parallel the lab component of the enterprise experience are included in Figure 6-4. These topics provide a skeletal synthesis of what is included in most manufacturing textbooks. The major topics appear somewhat universal while the subtopics are dynamic.

## The Automation/Control/Computer Applications Experience

Automation and computers have had a dramatic impact upon manufacturing, the people employed in the manufacturing sector, and society in general. These modern technologies have improved productivity, increased quality, raised the level of skills required in the workplace, and displaced countless workers who do not possess the needed skills. In technology education, the past decade has witnessed the development of computer software and benchtop systems whereby students can experience computer-numerical controlled machining, robotic systems, inspection with machine vision, and automated warehousing systems. Computer-assisted drafting (CAD) systems have become more sophisticated and at the same time easier to learn and use. What's more, they have become more affordable to educational institutions.

This new hardware and software has provided a means to teach content that otherwise would have been impossible to include except on a purely cognitive level. It has breathed a new sense of excitement into the profession and into the students that enroll in manufacturing courses.

Yet these advancements have not come without challenges. Because of the "bottom line" orientation of industry, the cost of upgrading to new equipment and software can be justified in real dollars. For example, if a new version of CAD software can reduce the number of steps that the drafter must perform to prepare a drawing, the new equipment can be written off in a relatively short period of time. The savings ultimately reduces the cost of doing business and becomes a key element in global competitiveness.

Expenditures in education, on the other hand, do not parallel a production model. It is much more difficult to justify new technology within the context of technology education even if it helps students learn more thoroughly, quickly, and effectively. A modern educational enterprise cannot afford to invest large amounts of resources in a technology that is risky in its propensity to become obsolete. Unlike the equipment of the past which had a useful lifespan of twenty years or more with little maintenance, today's equipment requires continual upgrading (often with sizable costs involved).

Organization of Manufacturing Industries

The major subdivisions of manufacturing and their responsibilities. Private versus corporate ownership. Global manufacturing.

Research and Development

Ideation including invention, innovation, and adaptation. Design and product development. Designing for assembly. Economic analysis.

Management and Personnel

Management theory. Hiring, firing, and training practices. Labor unions. Personnel development. Human relations and rights. Resolving conflict.

Finance and Accounting

Raising capital. Purchasing procedures. Bookkeeping and accounting. Direct vs. indirect costs. Break-even point. Profits. Closing the enterprise.

Marketing

Establishing company identity. Conducting market research. Advertising, sales, and distribution.

Product Packaging

Functions of packaging: protection, communication, containment, and utility. Designing and producing packages. Packaging and the environment.

Production Planning

Assessing resources. Flow process and operation process charts. Transportation within the plant. Storage.

Production Tooling

Types of tooling: jigs, fixtures, and templates. Tooling design and construction principles. Ergonomics and safety. Critical dimensions.

Plant Organization

Plant layout methods: process layout versus product layout. Custom, job shop, line production, and continuous production systems.

Quality Assurance

Tolerances and their relationship to costs. Developing a quality assurance system. Designing inspection devices. The "zero defect" concept.

Teaching Manufacturing

Safety. Designing manufacturing activities for various age levels. Managing students in a manufacturing class. Equipment and materials.

*Figure 6-4: Topics addressed in a manufacturing enterprise course.*

And the vocational education argument of duplicating the equipment of industry within the school is not very defensible, if it ever was, since technology education is clearly general education and is not intended to directly prepare production workers. As a result, the resources in technology education are often a step behind or a little less “flashy.” In brief, the preparation of manufacturing teachers with the latest technology remains a major challenge.

There is the risk of spending large amounts of time teaching collegiate students how a particular system works or how to use a particular piece of software, only to find that such learnings are rendered useless. This issue was delineated by Zuga and Bjorkquist (1989). Common applications such as word processing, simulation, and industrial control will likely withstand the test of time, while specific packages may disappear with the development of new products and new approaches.

Some questions need to be asked as one attempts to integrate computers and automation in technology education and, specifically, manufacturing education. Will the investment in time and energy to learn the new software or new technology produce an immediate return in terms of learning, new insights, or higher orders of problem solving? Is the software or new technology the means to greater efficiency and higher levels of thinking, or is it really an end in itself? Are the learning outcomes that result from the new equipment or software essential for all learners; that is, are the outcomes truly general education? Are the costs justified in terms of the learning outcomes? Are the learning experiences transferable and implementable in some manner in the public schools? Is there evidence that the concepts being taught are going to be adapted in widespread practice in the real world?

Recognizing the risk by virtue of the aforementioned questions, Figure 6-5 presents a hierarchy of computer-related experiences for the manufacturing teacher. The lowest level (i.e., General) includes word processing, spreadsheets, and databases. These are often the backbone of computer literacy courses offered by computer science or similar departments as service courses for the university at large. A key element in this level of the hierarchy is that the learners recognize and have the opportunity to apply these tools within actual manufacturing situations. This requires instructors of manufacturing-related courses to be cognizant of what the students have learned through their previous computer-related coursework and suggest or require its application in manufacturing. The application of word processing is self-evident. Spreadsheet software, for example, can be used to keep track of expenses, prepare break-even charts, and perform statistical analysis on product quality. Database programs can be used to keep track of personnel, maintain inventory, and prepare multiple forms.

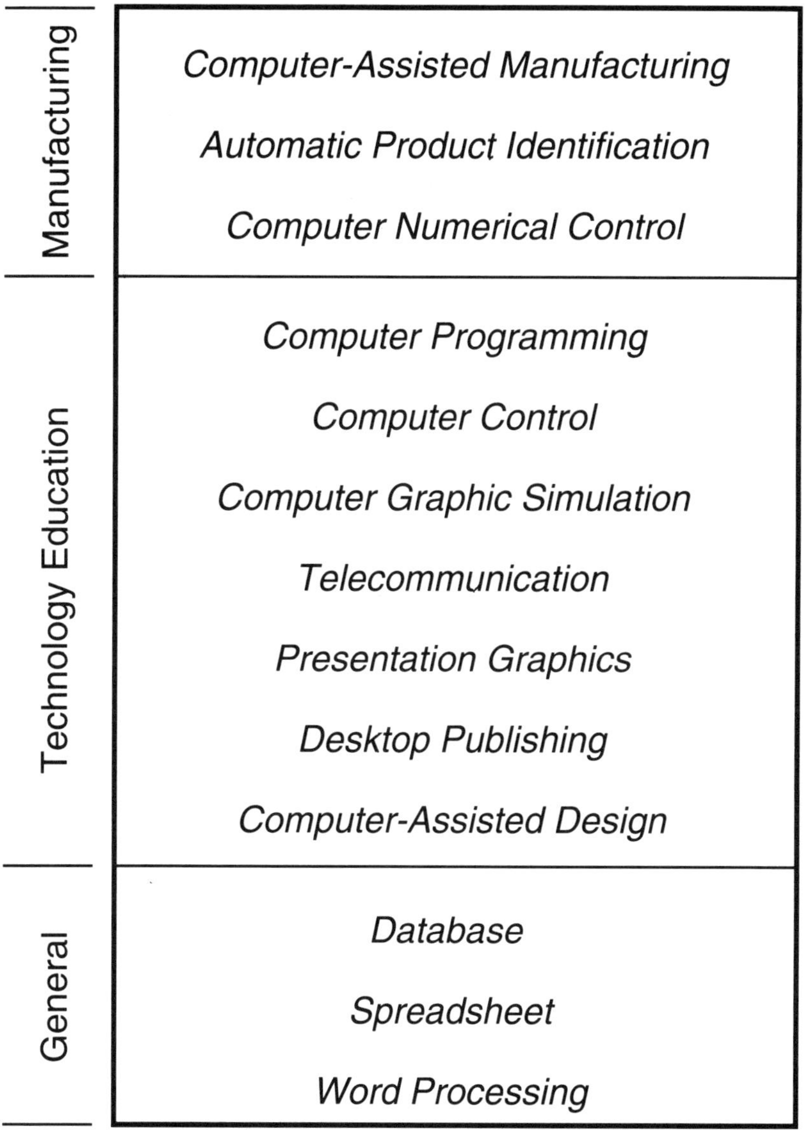

| | |
|---|---|
| Manufacturing | Computer-Assisted Manufacturing |
| | Automatic Product Identification |
| | Computer Numerical Control |
| Technology Education | Computer Programming |
| | Computer Control |
| | Computer Graphic Simulation |
| | Telecommunication |
| | Presentation Graphics |
| | Desktop Publishing |
| | Computer-Assisted Design |
| General | Database |
| | Spreadsheet |
| | Word Processing |

*Figure 6-5: Hierarchy of computer-related experiences for manufacturing.*

The Technology Education portion of the hierarchy (refer to Figure 6-5) includes experiences pertinent to technology education in general. At the lowest level in the hierarchy is computer-assisted design (CAD). Just as drafting was the "language of industry" in an earlier era, representing ideas graphically is the "language of technology" today. Desktop publishing is similarly important in the communications area; it is a new tool that has replaced many typesetting, paste-up, and printing functions. Presentation graphics have become the modus operandi in business and industry. There are numerous applications important in the manufacturing classroom including the presentation of product ideas, preparing company logos, and graphing the sales performance of the manufacturing enterprise throughout the course.

Telecommunications has become a matter of habit in most higher education institutions with electronic mail systems like BITNET and Internet. Public school teachers are increasingly becoming part of national and international computer communication systems. For example, a system called Virginia's Public Education Network (Virginia's PEN) was established in 1991. It consists of a computer system accessible by any teacher in the state with a personal computer and a modem. A toll-free number is available so that every teacher in the state can communicate with other teachers and professors with no long distance charges. This provides many opportunities for manufacturing teachers to communicate with one another and engage in global manufacturing projects. It also provides the means for students at both the public school and university level to communicate with one another and with other teachers.

All pre-service teachers should be literate about telecommunications. But like any new technology, there has to be a context and reason for its use. In manufacturing, students and teachers can share information, exchange CAD drawings for new product designs, and consult experts for the solution of technical and managerial problems. Telecommunications also provides an excellent means for simulating global manufacturing projects whereby parts for a product are subcontracted to students in distant schools.

Computer simulation is another technology that offers huge potential in the manufacturing education environment. It enables routine activities and problems from the production sector to be represented on the screen and allows opportunities for learners to interact with the simulation. Simulation programs exist whereby the user can design and program a workcell or simulate an entire production system on the computer screen. Unfortunately, the cost of such sophisticated software is prohibitive at the present time. On the other hand, simulations such as *The Factory* (Sunburst Communications) provide middle school students with the opportunity to design a simple production system and assess how it performs in the making of individual parts. Because of the three-dimensional nature of technology education, simulation programs are bound to play an increasingly important role in the instructional media in the next decade, particularly in manufacturing courses.

Computer data acquisition is another area which has pertinent applications to the collegiate manufacturing sequence. It involves using the computer as an input system, collecting physical, electronic, or mechanical data. For example, force and displacement sensors enable a learner to efficiently collect accurate data regarding the strength of materials. With linear measurement sensors, immediate feedback can be obtained about part tolerances, leading naturally to statistical process control. At a simpler level, the number of products passing a certain point in the production line can be counted and monitored using switches or light sensors.

In contrast to computer data acquisition, computer *control* uses the computer as an output device. Common applications involve triggering solenoids connected to pneumatic cylinders to clamp parts, feeding a drill into a piece of stock, or controlling the starting and stopping of a conveyor belt. Inexpensive interfaces, conceptually simple, are available which allow students to integrate computer control into a manufacturing system without prerequisite coursework. By using data acquisition together with computer control, students can design simplified versions of the most sophisticated automation systems used in manufacturing today.

At the highest level of the computer experiences in the collegiate technology education portion of the hierarchy, shown in Figure 6-5, is computer programming. When computers were first affordably made available, there was little else one could teach except programming since very little application software was available at the time. A generation of students who had some programming experience has come and gone. Most computer experience now is with software applications (LaPorte, 1988). However, programming allows the user to gain a significant amount of control and power over the computer. Students who have programming skills can develop custom record-keeping systems, time study applications, and graphics software that pertain explicitly to a particular production system.

There are three computer experiences that are unique to manufacturing in the computer experience hierarchy. While they are destined to change, they continue to be the mainstays in the modern world of manufacturing. One is computer-numerical control (CNC). In many respects the term has outgrown its usefulness—numbers are at the heart of all computer systems. However, CNC technology evolved from an earlier era when the computer did not play a significant part in the technology. Students should have experiences with programming a machine to produce a specified part. A system called "G-Codes" remains a common language to achieve this end. At the same time, the student should recognize that the time that it takes to conceive the G-Code program may not be an efficient use of human resources.

Virtually every product that appears in the marketplace is identified with some form of product identification mark, commonly called a "bar-code." Since there are several systems in use at the present time, the preferred term for this technology is "automatic product identification" or API. Because of its predominance in business and industry (particularly in manufacturing), it is included in the hierarchy of recommended learning experiences for prospective manufacturing teachers. For example, larger retail stores not only avoid the updating of price tags on each item as prices fluctuate, but they also maintain exact inventory of products through API. Orders for new inventory can be automatically generated by a computer and sent to a central warehouse with no human intervention.

Inexpensive bar code readers and bar code generation software are well within the budget of manufacturing educators. In a laboratory situation, API can be used to keep track of personnel, inventory, and products as they move through the production system.

Finally, prospective manufacturing teachers should become knowledgeable about computer-aided manufacturing (CAM). Though the term has undergone a metamorphosis of meanings in recent times, it at least involves the connection of CAD, CNC, and related technologies. More specifically, the computer file of a CAD part drawing can be precessed so that it yields the instructions necessary for the automated machining of the part. Software to accomplish such a feat is increasingly becoming affordable by the educational market. Similar to CNC machining, there are some issues involved in this technology which will be discussed in the last section of this chapter.

On virtually a daily basis, new computer-based developments occur which have implications for manufacturing education. These include computer-integrated manufacturing (CIM), just-in-time (JIT) manufacturing, and flexible manufacturing systems (FMS). However, the means of integrating these newer concepts into the manufacturing program typically rely upon the creativity of members of the profession.

## MANUFACTURING EDUCATION AND THE PROFESSIONAL SEQUENCE

Prospective teachers participate in numerous professional courses designed to cover instructional methods, prepare media, address topics such as student discipline and safety instruction, and develop effective evaluation systems. This component has become an integral portion of the collegiate major in technology education for two reasons.

First, manufacturing technologies are dynamic and manufacturing teach-

ers will be required to continually update their instruction throughout the rest of their professional careers. As new manufacturing techniques and systems are introduced, public school teachers will find it necessary to alter their activities, media, and strategies. Second, due to the tremendous change in the context of manufacturing education, even experienced teachers have been challenged to upgrade their professional skills. Teachers that graduated from industrial arts programs decades ago have returned to the classroom to learn how to structure cooperative group activities, teach creative problem solving, and address instruction of design topics. Learning how to teach others to use an engine lathe involves a different set of instructional strategies as compared to addressing the consequences of automation and global quality standards in the modern classroom.
This section will review several of the important themes in the professional sequence at the college and university level. A broad array of topics will be discussed including textbook selection, interdisciplinary activities, and the contribution of student organizational activities in preparing future manufacturing teachers.

## Integration with Other Areas

Integration of subjects has received considerable attention in public school programs. The movement to a core curriculum in higher education shares similar objectives. That is, the walls between subjects and disciplines are often broken down or have at least become more transparent.

There are a number of opportunities for the integration of manufacturing with other clusters in technology education. Several of these have already been mentioned. Additional examples include using communication and transportation concepts, technical knowledge, and skills during manufacturing coursework. Pre-service manufacturing teachers should learn how to integrate the development of trademarks, videotapes, and related media in marketing new products being produced during the enterprise experience. Transportation courses can involve students in designing and building systems for the transport of parts from one operation to another within the production facility. Mechanisms and systems for loading/unloading parts and products can also lead to challenging assignments for students.

Manufacturing education at the collegiate level often involves intra-program activities. The integration is accomplished through “contracting” some of the work that needs to be done. Not only does this increase the opportunity for developing and improving personal communication skills, but it often gives students in other classes real world problems to solve. For example, at Virginia Tech a course titled Communication Technology requires the students to develop a complete promotional campaign for a product or service. The instructor has used products that the manufacturing classes have produced as the focus of the promotional campaign.

There are also many opportunities that go beyond the technology education program. Students in management science can work with manufacturing students to improve their management skills and management system. A manufacturing systems class could be used as a case study in management. Business students could be involved in setting up an accounting and record-keeping system for the manufacturing enterprise class. Marketing students could be involved in designing a promotional system. Engineering students could be called upon to critique the production system, using actual data instead of examples from a textbook. Students of safety could analyze the manufacturing lab in terms of compliance with safety standards and practices. Art and industrial design students could become involved in the design of the product or the packaging as well as the design of a logotype for the company.

Finally, there are many opportunities to integrate the knowledge learned in other courses into the manufacturing sequence. Reference to history courses that include a study of the industrial revolution make both courses more relevant. The application of statistics and other areas of mathematics have immediate relevance in manufacturing. Physics principles can be applied to many of the problem-solving challenges that students face in a manufacturing class.

It is often difficult to integrate disciplines at the secondary level but it is even more so in higher education. It is surprising how autonomous and cloistered university departments behave. It is also surprising how little faculty members know about the courses they are requiring their students to take, not only across the university but even within the technology education program itself.

Therefore, there are numerous opportunities to enrich the manufacturing experience through activities with other areas. But, these activities require time, planning, and perhaps some additional resources. Some of the resources needed can be found within the students themselves as they are usually willing to do "what it takes" to make exciting things happen. The dividends for the professional *and* technical/technological sequence are perhaps limitless.

## Instructional Materials and the Preparation of Manufacturing Teachers

Today, manufacturing textbooks are in ample supply. However, as is true with technology education in general, there is a void in those available at the teacher education level. As a result, teacher educators have two options.

One is to choose texts that are intended for the secondary school level and then supplement them with assignments and readings that are appropriate for collegiate level students. The other option is to use textbooks suitable for industrial or engineering technology students. The trade-off with the latter option is that such texts are often far too specialized, may require prerequisite knowledge that prospective manufacturing teachers do not have or need, and may not present manufacturing in the context of a total industrial enterprise. What's more, they do not usually include the details of how to setup and operate an actual production system. In view of this, the best option is the first suggestion—use secondary texts and supplement them. Even though they are not generally within the reading level of college students, at least they provide a direct experience with a typical text that might be used when the manufacturing teacher begins their teaching career.

Visual media suffer from the same split between engineering levels and secondary levels. However, the problem is not as great since most media can be selectively assembled by the teacher and they do not hold the same significance in the course as does the textbook. Media specific to manufacturing education can be obtained through the regular suppliers of instructional resources for technology education. These vendors advertise regularly in professional journals and they are present at regional and international conferences.

Media and other resources on the latest developments in manufacturing are also available from the Society of Manufacturing Engineers (SME). This organization has been an enduring supporter of manufacturing programs in the public schools and provided considerable support to the Industrial Arts Curriculum Project World of Manufacturing program (Ray, W.E. & Lux, D.G., 1971). Several additional resources are available from the Center for Implementing Technology Education (CITE) at Ball State University. The CITE project involves the development and dissemination of action-based activities, curriculum guides, and related media.

The national Public Broadcasting System (PBS) and the Discovery Channel are excellent sources of current manufacturing information through video media. Copies of program tapes are usually available at a nominal charge at addresses listed at the end of the program. Teachers can also videotape segments with a VCR. However, care must be taken to assure that the use of videotaped features adhere to copyright laws.

Future manufacturing teachers should be taught the skills of preparing their own, custom media as well. Video cameras can be used to capture the action in a tour of a nearby manufacturing plant. Inexpensive copy stands and lenses enable the pre-service teacher to capture images from technical literature and trade journals. Multimedia presentations can be prepared to illustrate manufacturing concepts while allowing the learners to move through the material at their own pace.

Computer software, especially computer simulations, offer a great potential in manufacturing education. Unfortunately, there is a dearth of computer software that has been developed specifically for technology education and manufacturing education. There are plenty of general software available (such as those mentioned earlier), however, there are few that currently address modern manufacturing concepts. What exists often focuses upon drill and practice in nomenclature and facts. Hopefully this will change in the near future.

Manufacturing students must also become aware of the current literature that is available, where it is available, and how to obtain it. Reading assignments and report requirements are essential to reach this goal. Several trade journals are available at no charge to manufacturing teachers in the public schools. In addition, back issues may be provided by local industries through an "adopt-a-school" type of program.

Media can play a very important role in bridging the gap between the school laboratory and the real world of manufacturing. Manufacturing teachers should obtain the skills to properly obtain it, use it, and prepare it.

## Professionalism and the Aspiring Teacher

Since manufacturing technology is changing virtually day-by-day, one of the formidable problems facing the manufacturing teacher is keeping up with new developments. Perhaps most significant is the example set by the manufacturing teacher educator to the aspiring teacher. If the manufacturing course that the teacher educator provides is up-to-date and an overt effort is made to keep it that way, this is likely to have a positive effect on the student. Regardless of career or trade, the concept of "professionalism" is important in today's society.

Prospective teachers of manufacturing must be informed about the various professional organizations that serve the needs of the manufacturing community as well as the technology education profession. The Society of Manufacturing Engineers (Dearborn, MI) has become increasingly interested in supporting manufacturing experiences at the secondary level and has supported a number of events for technology education profession. Student chapters of SME are active on many university campuses. Though often directed by engineering students and faculty, they welcome membership from technology education students as well.

The Technology Education Collegiate Association (TECA), an affiliate of the International Technology Education Association (Reston, VA), has become increasingly stronger over the years. Each year regional and national contests are held whereby collegiate students compete as manufacturing teams to design and produce the best manufacturing system for a given product. Through such experiences, the aspiring teacher quickly sees the advantage of becoming active in professional activities.

Local TECA chapters often apply what they have learned, or are learning, in their manufacturing experience to producing products for sale as a fund-raising project to support their activities. These projects may be spin-offs of the manufacturing class or they may be independent of it.

Feeling a part of something positive and exciting is perhaps the most important outcome of professionalism. Through the example of their professors, through being current in the developments in manufacturing, and through involvement in student organizations like TECA, the aspiring teacher can achieve this outcome.

Public school teachers control the manufacturing curriculum they deliver. To decide whether the educational experience is positive or negative will be their decision for they are the ones that will make it happen or will prevent it. Beyond a solid grounding in the fundamentals of manufacturing, the best tools with which we can equip future teachers are resourcefulness, adaptability, and the ability to make rational decisions. As admirable as these qualities may be, perhaps the best means to teach them effectively is to provide a positive role model.

# A LOOK TO THE FUTURE

Many challenges face the manufacturing teacher of the future. Foremost is the viability of teaching manufacturing in view of the declining employment in the goods-producing economy. Fewer and fewer people are expected to be employed in the manufacturing sector in the future. Therefore, why bother to teach it at all?

Such an argument is perhaps defensible for vocational educators. But technology education, though it can be supportive of vocational education, is clearly general education and is becoming more so each year. Consumption of manufactured products will most certainly not decrease. Likewise, the production of manufactured goods will not decrease either. It is simply a matter of producing products with fewer people. An understanding of how goods are produced is equally important whether they are produced by thousands of employees or by a single worker. If manufacturing is truly general education and a vital experience for everyone, then the number of people employed has little bearing.

One trend receiving increasing attention in technology education is that of the biotechnologies. Savage and Sterry (1991) have suggested biotechnology be an additional content organizer for the field. However, biotechnology is clearly defensible as a part of manufacturing. For example, Smith (1988) defines biotechnology as the "use of microbial, animal, or plant enzymes to synthesize, break down or transform materials" (p.1). Including biotechnology as a part of manufacturing would allow the profession to nurture its development and determine its feasibility over time.

With a solid argument for the continuance of manufacturing education in hand, the challenge turns to what should be taught to the prospective teacher in the future. In the earlier era of industrial arts, consistent attention was paid to psychomotor skills. Dugger (1980) reminds us that "skill in the use of tools and machines" was considered to be the most important goal of industrial arts just over a decade ago. However, in a recent study on the ideal emphasis on goals for technology education in Virginia by Yu (1991), this goal was ranked only seventh in importance among technology teachers, and only tenth in importance among teacher educators and state leaders. Furthermore, there seems to be little attention in current technology education literature about the importance of psychomotor skills.

If one assumes that there is, indeed, a de-emphasis on the importance of teaching psychomotor skills, then it suggests that new issues must be addressed. Numerous technological innovations and enhanced educational methodologies will dominate future collegiate programs.

For instance, the area of manufacturing design is encountering tremendous change. With software that has been available for several years now, it is possible to take a wire-frame model developed on a CAD system and convert it into the instructions necessary to operate a CNC machine and produce the part. Once the wire-frame model has been produced with CAD, the only psychomotor skills necessary are mounting the stock in the machine and typing the necessary commands on the keyboard.

The ultimate question, reflecting the general education purposes to which our programs aspire, involves "What does everyone need to know and be able to do to effectively live in our technological society?" Does everyone need to know G-codes, conventional machining, or how to connect the CAD system with a CNC machine? Remember, the purpose is not to prepare technicians vocationally, but rather to prepare everyone regardless of their career aspirations. This creates an entirely new perspective on what it is we are doing, should be doing, and how we address the needs of a population that is to become technologically literate.

If one takes a look at a sample of some of the new technologies currently in their infancy, the challenge (as well as the potential) becomes even greater. One of these emerging technologies is what is often called "stereo lithography" (or perhaps "rapid prototyping"). Using a photosensitive liquid polymer, a laser beam polymerizes successive layers of the polymer, thereby producing a three-dimensional plastic part from a wire-frame description of the part. Stereo lithography overcomes some limitations to the CAD-CAM connection. For example, if a part requires lathe turning as well as milling, then the CAM system is typically confused. The same is true of parts that can only be produced by precision casting.

At this time, the process is limited to a particular polymer with a narrow range of properties. However, the process will no doubt be expanded to a wider range of materials in the near future. Though it requires several conceptual jumps, it is not inconceivable that the science fiction concept of "beam me up, Scotty" from Star Trek, at least as it pertains to products, could become reality within the careers of the technology teachers we are now preparing. That is, products could be produced at one site and "cloned" at a remote site, effectively tele-transported. The technology is not complex and therefore affordable educational versions of the equipment could no doubt appear on the market in the near future.

A second technology that will no doubt have an impact upon manufacturing education is "virtual reality." Using sophisticated video equipment, a person is presented with an audio-visual context that produces a very realistic, multi-sensory experience. What's more, the viewer (i.e., the "participant") has control over the situation so that he or she actually interacts with the situation. Instead of passively viewing a movie, for example, the participant determines where to go and what to do.

Using virtual reality, students could manipulate an entire manufacturing system. Tools and equipment could be moved about and tested. Even learning how to use a particular piece of equipment could be realistically simulated through virtual reality. Already, the State of Texas is using an elemental form of virtual reality to teach pre-service teachers how to deal with discipline and classroom management problems (Rohrbough, 1991).

This new technology that is now on the horizon of education has many implications for the future. Indeed, it could challenge the very nature of what constitutes a hands-on experience. Further, it is conceivable that students in the future will be able to achieve many of the goals and objectives we now espouse in manufacturing education exclusively through an experience with computers and related technology.

## SUMMARY

This chapter reviewed the nature of manufacturing education at the college and university level. At this point, the manufacturing program involves the preparation of knowledgeable, laboratory-competent teachers. Three elements of the collegiate program were described including general education requirements, the technical/technological component, and the professional sequence.

Courses in the technical/technological block are often taken with majors from other technology and engineering programs. The Holmes Group suggests that a full four-year program in a technology-based discipline precede a graduate degree in technology education. Unfortunately, this technique fails to allow students to realize the true context of modern manufacturing education. The "Hour Glass" model presented in the chapter seems to be the best alternative for preparing future manufacturing teachers.

The professional sequence involves methods courses and student teaching. In light of new demands for technological literacy among all students, the emphasis on effective teaching methods has increased. Future teachers must know how to teach problem solving, design, organize and supervise team assignments, etc. Finally, professionalism is enhanced through student organizational (club) activities plus educational and trade association activities.

# REFERENCES


Carnegie Task Force on Teaching as a Profession. (1986). *A nation prepared: Teachers for the 21st century*. New York: Carnegie Forum on Education and the Economy, Carnegie Corporation.

Dugger, W.E., et al. (1980). *Report of survey data: Standards for industrial arts program project*. Blacksburg, VA: Virginia Polytechnic Institute and State University.

Earl, A. W. (1960). *Experiments with materials and products of industry*. Bloomington, IL: McKnight.

Holmes Group. (1986). *Tomorrow's teachers: A report of the Holmes Group*. East Lansing, MI: Holmes Group.

Israel, E. N., & Wright, R. T., (Eds.). (1987). *Conducting technical research*. 36th Yearbook, Council on Technology Teacher Education, Mission Hills, CA: Glencoe.

LaPorte, J. E. (1988). *The use of selected advanced technologies in technology teacher education programs*. Paper presented at the annual conference of the International Technology Education Association, Council on Technology Teacher Education, Norfolk, VA (available from the ITEA Technology Bank).

Loucks-Horsley, S. (1990). *Elementary school science for the 90s*. Alexandria, VA: Association for Supervision and Curriculum Development.

Maley, D. F. (1988a). *Organizing the content—with emphasis on R & E programs*. Reston, VA: International Technology Education Association.

Maley, D. F. (1988b). *R & E program for the gifted and talented*. Reston, VA: International Technology Education Association.

Miller, C. D. (1991). *Programs to prepare teachers of technology education*. The Journal of Epsilon Pi Tau, *17*(1), 59-69.

Nelson, H. F. (1962). Implications for program. In D. E. Lux, (Ed.), *Essentials of Pre-service Education*, (pp. 169-178). 12th Yearbook of the American Council on Industrial Arts Teacher Education, Bloomington, IL: McKnight & McKnight.

Novotny, J. A., Thomasser, T. S., & Chapin, D. W. (1990). *Global manufacturing project*. Paper presented at the annual conference of the International Technology Education Association, Indianapolis, IN.

Ray, W. E., & Lux, D. G. (1971). *The world of manufacturing*. Bloomington, IL: McKnight.

Rohrbough, L. (1991, November 13). *Newsbytes*. General Electric Network Information Exchange (subscription telecommunication service).

Savage, E., & Sterry, J. (1991). *A conceptual framework for technology education*. Reston, VA: International Technology Education Association.

Smith, J. E. (1988). *Biotechnology* (2nd ed.). London: Edward Arnold Publishers.

Starkweather, K. (1991, November). ITEA/CTTE/NCATE accreditation: Its status and implications for the profession. In D. Householder (Ed.), *Technology Education and Technology Teacher Education*. Proceedings of the Mississippi Valley Teacher Education Conference.

Towers, E. R., Lux, D . G., & Ray, W. E. (1966). *A rationale and structure for industrial arts subject matter*. The Ohio State University Research Foundation.

Wright, R. T. (1990). *Manufacturing systems*. South Holland, IL: Goodheart-Willcox.

Yu, K. C. (1991). *A comparison of program goals emphasized in technology education among selected professionals in the State of Virginia*. Unpublished doctoral dissertation, Virginia Polytechnic Institute and State University, Blacksburg.

Zuga, K. F., & Bjorkquist, D. C. (1989, Fall). The search for excellence in technology education. *Journal of Technology Education*, *1*(1), 69-71.

CHAPTER 7

# Establishing The Manufacturing Teaching/Learning Environment

Dr. Jack W. Wescott (Assistant Professor)
*Ball State University, Muncie, IN*

The teacher of manufacturing in a technology education program must employ a wide variety of instructional strategies in order to be effective. The relationship between content and methodology greatly influences the selection of the proper instructional strategy. Few would disagree that the content of contemporary manufacturing programs is significantly different from the subject matter of previous decades. Among the differences, besides the advancing technological content, are newer instructional strategies designed to promote critical thinking and problem-solving skills. Therefore, a diverse set of instructional strategies must be implemented if effective learning is to take place.

A review of educational literature reveals a tremendous variety of teaching strategies. This chapter begins with a discussion of learning theories and the individual needs of manufacturing technology students. A specific model for the development of instructional strategies is also presented. Also, a mixture of contemporary instructional strategies appropriate for teaching manufacturing technology are addressed. The chapter concludes with a review of the evaluation and assessment process appropriate for the study of manufacturing.

## LEARNING THEORY AND STYLES

No matter how relevant the material or how enthusiastic the teacher is about the content, learners must be able and willing to learn. Therefore, concern for students must be the foundation of all planning efforts. In thinking about the learning needs of students in a manufacturing program, it is important to remember that individuals learn in different ways and that teachers have different teaching styles. In their research on learning styles, Silver and Hanson (1991) indicated that some people are more intuitive than others. They also noted that some individuals respond to the material delivered in a logical order; yet others learn better through problem solving. Students with various learning styles require different instructional strategies, and all students need to learn in more than one way. The more the teacher knows about the learning styles of individuals, the better he or she is able to plan a variety of instructional strategies appropriate for the content of manufacturing technology. Certainly, there will be a combination of learning styles present in every class. If a teacher cannot vary instruction to account for individual differences, some students will be left out of the instructional process.

## The Needs of Learners

Learning in a manufacturing environment typically takes place within a set of events called learning strategies. Even before these strategies are put into effect, however, the learner must have been prepared for learning. The learner needs to have a readiness for learning which has been established by a number of additional events that precede the instructional strategy.

Gagne (1970) identified three factors that comprise learning readiness including the student's: a) attentional sets, b) motivation, and c) developmental status. An attentional set is an internal state of mind which enables the learner to select stimuli appropriate to the learning experience. The term "attention" serves as a general name with several subsets in existence. For example, making specific responses, carrying out a sequence by following verbal directions, and exploring the environment have all been distinguished as separate activities of the attentional set. These pre-learning capabilities appear to belong to the more general category of cognitive strategies which govern an individual's information-gathering behavior.

The motivation of the student is a large and complex topic and attention should focus on those two kinds of motivations that are relevant to learning in the manufacturing environment. First, social motivation includes a need for affiliation, social approval, esteem, etc. Such motives are undoubtedly significant when considering the relationship between the teacher and young students. The second category of motivation is task mastery and achievement. In general, such motivation is enhanced by successful experiences (i.e., achieving the specific objectives of learning tasks).

A major technique used to analyze individual differences and learner readiness among students is the study of human growth and development. More specifically, the use of developmental tasks to identify the capabilities of individuals has proven to be a valuable asset to the selection of instructional strategies. A description of the significant role of developmental tasks in the educational process was described by Havinghurst (1974) when he stated:

> There are certain guideposts which are helpful in gaining an overall picture of growth and development. These guideposts we call "developmental tasks." The concepts of developmental tasks provide a framework within which we can organize our knowledge about human behavior and learn to apply this information in dealing with children in our schools. (p. 77)

The use of developmental tasks as a fundamental base for program development in manufacturing education is predicated upon the following assumptions:

- That the science of human growth and development has established the validity of "developmental tasks" as a phenomena associated with the normal growth of all individuals.
- That the developmental tasks should be used as a fundamental base for program or curriculum development.
- That activities may be structured to contribute to the individual's accomplishing of various tasks.
- That it is possible to integrate structured activities into the instructional program so that the student may pursue his/her developmental tasks while accomplishing the educational and social objectives of the program.

Maley (1977) identified a four-stage model for the implementation of developmental tasks within a program. The model suggests a series of analytical steps, moving from specific developmental tasks to a set of guidelines for technology education. A graphic representation of this model was adapted for the study of manufacturing and is presented in Figure 7-1.

# DEVELOPING THE INSTRUCTIONAL SYSTEM

If teaching involves the design of an environment to facilitate learning, then it is necessary for the teacher to make decisions that identify and organize content in a meaningful fashion. These key decisions are easier if completed in a systematic way. Teaching should be a planned activity based on continuous feedback relative to the teaching/learning process. A model,

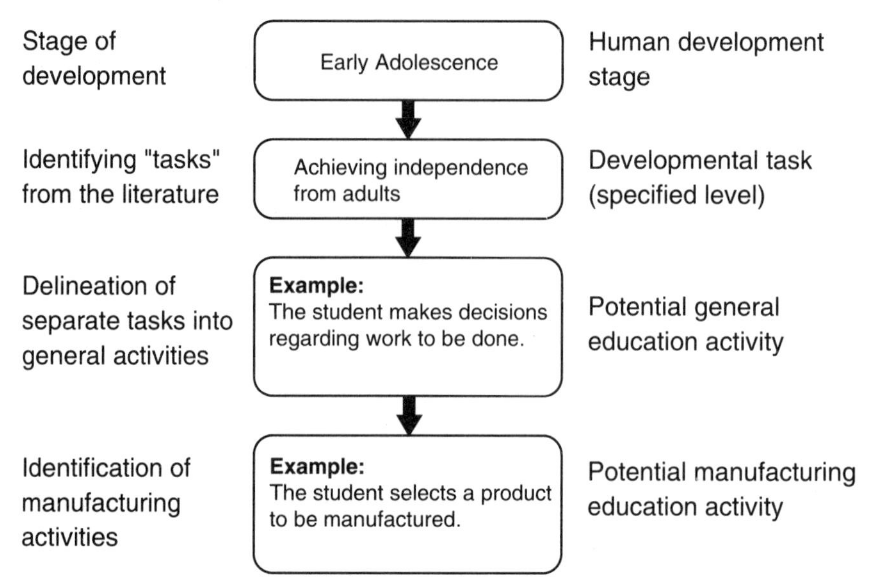


*Figure 7-1: The use of a four-stage model to identify and structure a simple manufacturing activity.*

illustrated in Figure 7-2, outlines the basic functions regarding the design and implementation of an instructional strategy. Although there are many such models which describe the identical process, this model has been selected because of its appropriateness to the manufacturing sequence in technology education programs. The model involves a series of distinct components. Each of these six components will be further described in order to provide a more intuitive understanding of the model.

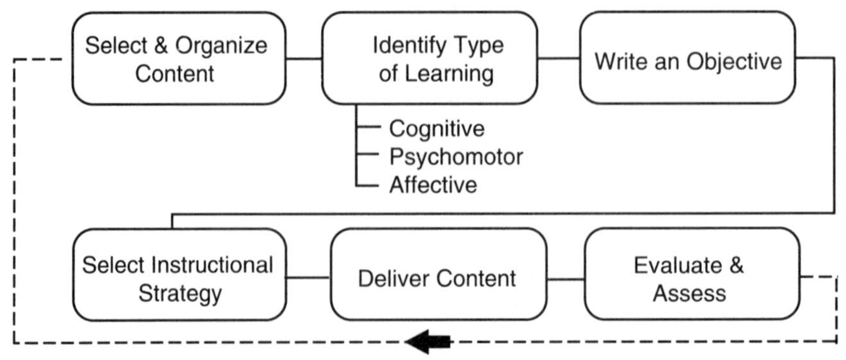


*Figure 7-2: Model for designing and implementing instructional strategies.*

## Selecting and Organizing Content

During this first stage, the challenge for the manufacturing technology teacher is to identify those topics that will be developed into a specific instructional strategy. The content for a manufacturing program is usually selected by the teacher based upon a planned curriculum or guide. For the purposes of our discussion, content relates to those facts, truths, principles, and mental processes related to manufacturing systems, design, materials, processes, management, and enterprises; and their consequences. Since manufacturing coursework is often divided into specific themes, each class presents a challenge to the teacher. The important themes related to manufacturing technology, and specific content for individual courses, has been described in earlier chapters.

There are three essential aspects related to the identification of content to be presented in a technology-based program: scope, focus, and sequence. Scope defines the breadth and range of the content to be covered. Focus determines what will be emphasized in the content, while the sequence specifies the order in which content will be arranged.

## Identifying Types of Learning

One of the unique characteristics of a manufacturing education program is its potential to present content in the three major domains of learning as identified by Bloom. The three types of learning are cognitive, affective, and psychomotor. Cognitive learning relates to the mental processing of information by the learner. The original cognitive theories of learning have been organized into a taxonomy, which is the basis for Figure 7-3. It is important to note that Bloom's taxonomy is not a statement of educational objectives. Rather, it is a system for classifying learning with respect to cognitive categories. The major significance of Bloom's taxonomy is that it serves to remind teachers that instruction should do more than promote the memorization of facts, details, and truths. An effective manufacturing technology program must include instructional strategies that are designed to challenge students to process information at higher levels.

The manufacturing sequence also has the potential to involve instruction in the affective learning domain. Learning in this domain concerns feelings and attitudes that individuals are expected to develop as a result of instructional activities. For example, the manufacturing student who works on a simulated assembly line may quickly develop an appreciation for monotonous work. Within the laboratory-centered manufacturing curriculum, students have the opportunity to experience many attitudes, feelings, and desires. Teachers often fail to include this type of learning in their

**Sample of Verbs in the Cognitive Domain**

| | |
|---|---|
| • Knowledge | to recall, repeat, recollect, memorize, and list |
| • Comprehension | to identify, recognize, and select |
| • Application | to use, solve, practice, reproduce, compare, and contrast |
| • Analysis | to investigate, separate, study, research, describe, and distinguish |
| • Synthesis | to combine, formulate, deduce, unite, assemble, and create |
| • Evaluation | to appraise, judge, assess, assign value to, accept, and reject |

*Figure 7-3: Examples of action verbs which operationalize the six levels of the cognitive learning (adapted from Gunther, Estes, and Schwab, 1990).*

**Sample of Verbs in the Affective Domain**

| | |
|---|---|
| • Receiving | to take in, listen, encounter, and be aware |
| • Responding | to react, reply, answer, and comply |
| • Valuing | to accept, reject, esteem, regard, and desire |
| • Organizing | to compare, to order, to prioritize |
| • Characterizing | to internalize, personalize, and demonstrate |

*Figure 7-4: Examples of action verbs which operationalize the five levels of the affective domain (adapted from Gunther, Estes, and Schwab, 1990).*

programs since it is difficult to plan and assess. The affective domain of learning is also organized into a taxonomy, as illustrated in Figure 7-4.

In the psychomotor domain, learning relates to mind-hand coordination. Although this domain has not received the attention of the cognitive and affective domains, it still influences a valuable outcome of an educational endeavor. The sole purpose of a manufacturing sequence in technology education is not the mastery of specific technical skills. However, students should be able to develop general psychomotor skills that are appropriate for the developmental stage of the individual, Figure 7-5.

## Sample of Verbs in the Psychomotor Domain

| | |
|---|---|
| • Readiness | willing, prepared, and watches |
| • Observation | attends and is interested |
| • Perception | senses, has a feel for, and is able |
| • Response | practices, imitates, and replicates |
| • Adaptation | masters, develops, and changes |

*Figure 7-5: Examples of action verbs which operationalize the five levels of the psychomotor domain (adapted from Gunther, Estes, and Schwab, 1990).*

## Writing Objectives

The purposes of instructional experiences are usually stated as objectives. The writing of objectives is an important part of the instructional design process for two reasons. First, objectives clearly identify the content of what is to be taught. Developing objectives is the initial step in the identification and organization of the subject matter to be presented. Second, the instructional objective assists the teacher in the identification of the type and level of learning as well as indirectly assisting with the selection of a meaningful instructional strategy. Instructional objectives also serve as a means of communication to various populations including students, teachers, parents, and administrators. Eventually, the objectives will play an integral role in the evaluation of the teaching-learning process.

There are many designs and suggestions for writing objectives. The most universally accepted (and probably the clearest) guidelines are identified by Mager (1962). Mager's instructional objectives have four elements which describe: a) who is performing the task, b) the desired terminal behavior of the student, c) the given conditions under which behavior will occur, and d) the level of achievement the learner will be asked to attain.

Manufacturing educators should address four general concerns when assessing the effectiveness of an instructional objective:

- Is the objective relevant?
- Is the objective clearly stated?
- Is the student behavior measurable?
- Does the objective include the four elements as identified by Mager?

In addition to these four concerns are several additional issues related to the use of objectives that are consistently expressed in the literature.

The first issue is that there is too much concentration on the lower levels of learning. For example, in the cognitive domain we tend to focus on behaviors such as "list, define, name, and select." This is perhaps due to the fact that it is easier to write objectives that are classified as one of the three lower domains. Likewise, objectives written at this level are easier to teach using direct learning methodologies such as formal lectures and demonstrations. Also, the evaluation of these objectives is easier through the use of simple evaluation procedures such as written tests and required laboratory assignments.

The two examples below illustrate how an objective for a manufacturing education program can be written for a specific behavior at different levels of cognitive development.

*Example #1:* Lower level objective
On a written test the student will correctly define the secondary process of separating.

*Example #2:* Upper level objective
Given various examples of processed materials, the student will correctly distinguish those that involved separation processes and explain the advantage of using each process.

The second issue relates to the perceived notion that instructional objectives do not allow for individual differences among students. Since it is virtually impossible to individualize learning outcomes in most manufacturing programs, teachers are forced to write objectives that address the needs of a particular group or class of students. A technique of accounting for individual differences is to write objectives with a level of achievement that allows for varying levels of abilities and backgrounds.

In summary, learning objectives describe the learning that is intended to take place as a result of instruction. In developing cognitive, affective, and psychomotor objectives, educators determine what should be included in the instructional design. Only when objectives are clearly defined can the teacher select appropriate instructional methods. Also, evaluation procedures written in conjunction with the objectives enable the teacher to better assess what intended outcomes have been achieved.

## Selecting an Instructional Strategy

Different types of learning outcomes require different instructional strategies. For example, the teacher who wishes to develop higher-order thinking skills must select an instructional strategy that is capable of accomplishing the desired outcome. As a result, it is recommended that one select an instructional strategy only after the content and type of learning have been identified and the objective(s) have been written. Although this appears to be a relatively simple principle, it is often neglected. From an instructional design standpoint, the major justification for breaking down and analyzing content is to assist in specifying the most appropriate instructional strategy.

## Delivering Content

Whenever a teacher of manufacturing technology plans an instructional strategy, they choose a means of organizing content from a range of options (given the real-life constraints of the laboratory setting). Once the decision is made on what is to be taught, the teacher must then decide how and when the material is to be taught. It is important to sequence instruction that is meaningful to the student—not the teacher. The "teach as I have been taught" approach to the organization of content is not always the most effective strategy for the learner. Traditionally, the two most common means of organizing content is through deductive and inductive organization.

In a deductive design, the strategy usually begins with the presentation of a generalization, a rule, or a concept. Students are then given specific examples along with facts associated with the generalization, concept, or rule. While moving from the general to the specific, students are encouraged to draw inferences and make predictions based on the examples.

In an inductive design, the students are first presented with the specific data and facts. Then gradually, through the process of investigation and reasoning, the students form the generalization, rule, or concept definition. The induction design is considered to be more conducive to stimulating the student thought process. However, deductive models can be very effective when used judiciously and sparingly in delivering complex content. Regardless of the system used, the sequence of content is important when designing an instructional strategy.

## Evaluating and Assessing

The distinction between assessment and evaluation is an important one. Evaluation involves a judgment of the quality of student outcomes measured against the learning objectives. In contrast, assessment is a deliberate appraisal of the effectiveness and nature of the teaching and learning that has taken place. It represents an action process in the teaching/learning process.

Determining the success of learning requires planned evaluation by both teacher and students. This evaluation process plays a key role in helping students develop as learners and in enhancing teachers' effectiveness. Evaluation plays a very important role in the manufacturing technology program. It is an integral part of the instructional design and provides information which serves as a basis for a variety of educational decisions. The main emphasis in educational evaluation, however, is the individual and the learning process.

From an educational viewpoint, Gronlund (1982) defines evaluation as a systematic process of determining the extent to which educational objectives are achieved by learners. It is important to note that this definition implies a systematic process so may not always include casual, random observation of students. Gronlund also notes that while the interdependent nature of teaching and learning is beyond dispute, the interdependent nature of teaching/learning and evaluation is often less clear. The evaluation process includes four fundamental steps:

1. Identifying and defining of student outcomes in terms of desired changes in behavioral terms.
2. Planning and directing learning experiences in conjunction with the objectives.
3. Assessing student progress toward stated objectives.
4. Using the results of evaluation to improve the teaching/learning process.

It should be noted that the process of evaluating student progress in a manufacturing program can be extremely challenging because of the activity-oriented nature of the content, the variety of types, the different levels of learning, and the use of complex strategies such as cooperative learning, problem solving, and conceptual learning. Due to these variables, alternative evaluation techniques are suggested.

# INSTRUCTIONAL STRATEGIES

The need to teach higher order thinking skills is not a recent one. For instance, Raths, Jonas, Rothstein, and Wassermann (1967) identified the lack of emphasis on thinking in schools several decades ago. They noted that memorization, homework, and a quiet classroom were often rewarded while "... inquiry, reflection and the consideration for alternatives were frowned upon" (p. 3). The increasing emphasis upon teaching higher order thinking skills adds a new dimension for the teacher of a manufacturing technology program. Now, instead of solely concentrating on accumulating masses of detailed information, teachers must also focus upon instructional strategies that are designed to enhance the full understanding of concepts and promote creative problem-solving skills. In practice, it is hoped these new strategies will help "teach students how to learn."

Traditional teacher-directed strategies such as the lecture and demonstration should be supplemented in order to encourage students to think logically, analyze, compare, question, and evaluate. This section presents an overview of selected strategies that are appropriate for learning in a contemporary manufacturing technology education sequence. These strategies include concept learning, cooperative learning, an interdisciplinary approach, and problem solving.

## Conceptual Learning

During the past three decades, there has been a significant attempt to identify and organize the content of modern manufacturing technology. These efforts have resulted, for all practical purposes, in an established curriculum (or set of organizers). However, minimal attention has been directed toward the specific instructional strategies that are appropriate for teaching the contemporary content of manufacturing technology. In order to address this issue, it is necessary to make two major assumptions.

First, it is assumed that simple demonstrations and related lessons should be supplemented with additional strategies designed to teach higher levels of learning. That is, manufacturing teachers need to make a clear distinction between content which is "conceptually based" versus "operationally based" and select a strategy that is appropriate for the desired learning outcome.

Second is the premise that a substantial portion of manufacturing content is conceptually oriented. This assumption is supported in section three of *Technology Eduction: A Perspective on Implementation* (1985) which established a criteria for technology education programs. This criteria supports the extensive use of terms such as "concept," "technological concepts," and "abstract concepts." Based upon these assumptions, it seems appropriate to address the issue of conceptual learning and its implications for manufacturing education programs. Additionally, Mitzel (1982) indicated the importance of concept learning strategies for classroom instruction when he stated: "Concepts are both the key building blocks of instructional materials such as textbooks and the main element of a learner's cognitive structure. Unfortunately, teaching of concepts often receives less emphasis than teaching of isolated facts and skills" (p. 896).

A review of research associated to the teaching of concepts is useful in the development of instructional strategies. Even though the majority of research in this area has been conducted by educational psychologists, Israel (1972) identified guidelines for using verbal and nonverbal instructional media to help develop an abstract understanding of technical concepts in junior and senior high programs. A summary of the research relative to teaching concepts identifies three major components: a) concept definition and identification of attributes, b) presentation of examples and non-examples, and c) establishment of hierarchial relations.

*What is Conceptual Learning?* A technical or technological "concept" is an attempt to apply the learning theories of the educational psychologist with a body of content which is technical/technological in nature. Since manufacturing involves simple techniques (mechanical, pneumatic, etc.) and managed activities (the human condition is represented by the *-ology* segment of the term, technology), there are three basic reasons for why manufacturing educators should provide conceptual instruction:

1. To provide an explanation of a **technical/technological process**. A *process* is defined as a series of actions or operations performed to provide a particular result. Examples of technical processes are such concepts as separating, casting, and molding, while technological processes include community planning and inspection.

2. To provide an understanding of a **technical/technological quality**. The term *quality* implies a characteristic that belongs to the essential nature of a material. Technical qualities include hardness, tensile strength, and impact strength. Technological qualities refer to the nature of products and services available to the average consumer.

3. To provide an understanding of a **technical/technological abstraction**. The term *abstraction* is an action or process that lacks a visual image. The concepts of electron flow, covalent bonding, and capacitance are examples of technical abstractions. Quality assurance and global manufacturing can be perceived as technological abstraction.

Much of the instruction in a manufacturing education program consists of learning concepts in the form of a process, quality or abstraction. These three components form the basis for the design and implementation of instructional strategies. Figure 7-6 illustrates several common manufacturing processes organized in technical concepts as identified by Wright (1991).

*Planning for Conceptual Learning.* Figure 7-7 provides a flow chart for the planning process for teaching a manufacturing concept. It should be noted that the secondary processing techniques illustrated in Figure 7-7 are

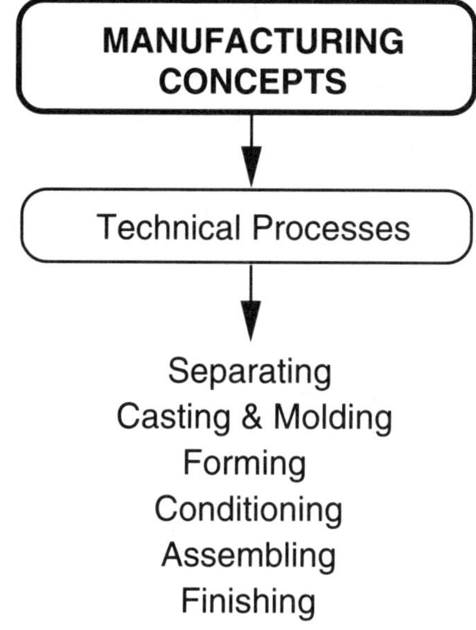


*Figure 7-6: Technical concepts of material processing.*

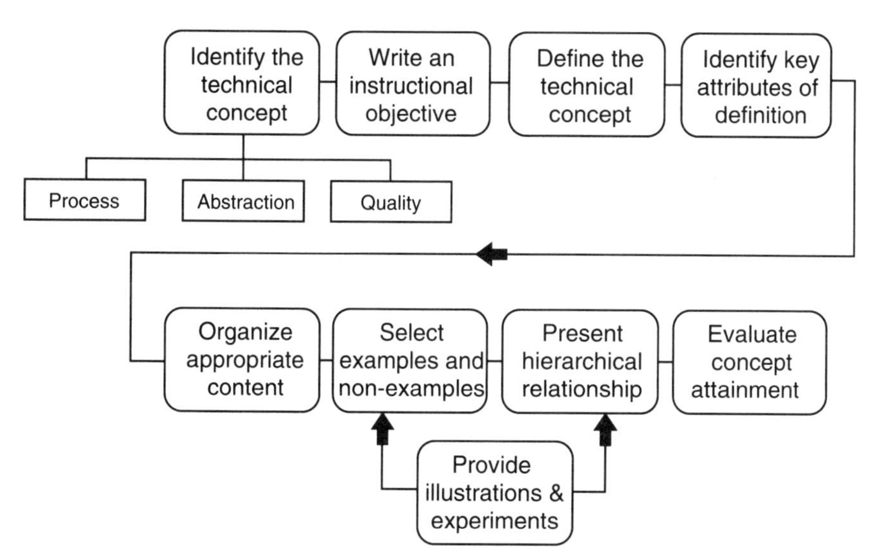


*Figure 7-7: A model to support the understanding of the hierarchical relationships between the attributes of a concept.*

intended to be generic and not inherent to a specific material. To begin the instructional planning process, it is necessary to state an instructional objective utilizing behavioral terms. After an objective has been established it is necessary to define the concept and identify the relevant attributes. This step is important since the definition identifies the content to be included in the instructional process. The definition, when combined with its attributes, forms the content of the lesson. This is based upon the assumption that an understanding of the attributes and their relationship to each other will provide the learner with an intuitive understanding of the process, quality, or abstraction.

Research shows that simply providing the learner with a definition of a concept does not ensure conceptual attainment. Specifically, Klausmerrer (1976) indicated that if a definition is provided without the presentation of examples and non-examples, the learner may simply memorize a series of verbal associations. Finally, the learner should be provided a graphic representation of the concept. This graphic model serves as a mental map for the learner and assists in understanding the hierarchial relationships between the attributes of the concept.

Finally, it should be made apparent that conceptual learning in manufacturing programs is not intended to replace or make obsolete traditional instructional strategies. However, the successful implementation of manufacturing programs is at least partially contingent upon the implementation of instructional strategies that address the issues of conceptual learning.

## Cooperative Learning

Due to the nature of the content and the action-centered, laboratory activities, cooperative learning is an effective strategy for teaching manufacturing technology. Cooperative learning is a teaching strategy involving student participation in group learning activities that promote positive interaction. Research has indicated that cooperative learning promotes academic achievement, is relatively easy to implement, and is not expensive. Students with improved behavior and attendance are some of the benefits of cooperative learning (Slavin, 1987). They also increased their opinion of school, in general.

*Advantages of Cooperative Learning.* Components of the cooperative learning process as described by Johnson, et al. (1984) are complimentary to the goals of manufacturing technology programs. For example, well-constructed cooperative learning tasks involve positive interdependence on others and individual accountability. To work successfully in a cooperative learning team, students must also master interpersonal skills essential for the group to accomplish its tasks. Cooperative learning has also been shown to improve relationships among students from different ethnic backgrounds. Slavin (1980) states that cooperative learning methods embody cooperative, equal status interaction between students of various ethnic backgrounds. For most students, teaching has traditionally stressed competition and individual learning. When students are given cooperative tasks, learning is assessed individually and rewards are often given on the basis of the group's performance.

Henak (1988) identified five reasons for why technology teachers should use cooperative learning strategies including:

1. Achieve higher levels of understanding of technological systems.
2. Develop higher level learning and critical thinking skills.
3. Contribute more to the well-being of others.
4. Develop higher levels of social responsibility.
5. Become effective participants in cooperative human action.

*Planning for Cooperative Learning.* Cooperative learning is not new to the manufacturing sequence. Due to the nature of the content and laboratory approach of the program, cooperative learning has traditionally served as an effective instructional strategy. The organization of students into small teams to complete tasks related to such areas as research and development, market analysis, product design, and production planning are examples of cooperative learning. In addition, nearly all of the activities in an enterprise class involve the cooperative efforts of the entire class.

Planning for cooperative learning involves the establishment of both process and content goals. Process goals refer to the type of social and emotional values that the teacher wishes to develop. Examples of process goals include courtesy, self-esteem, respect for others, acceptance of individual differences, and enjoyment in learning. These are examples of intangibles that determine the classroom atmosphere. In contrast, content goals refer specifically to the subject matter of the curriculum. They generally refer to the facts, truths, principles, and processes of manufacturing technologies. These goals are typically prescriptive and can be viewed as a statement of overall intention on the part of the teacher.

The implementation of cooperative learning suggests that learning involves a process, and one way to communicate this process is to consider the group learning model generated by Reid, et al. (1989). This five-stage model is explained in Figure 7-8.

The model by Reid has significant applications for teaching manufacturing. For example, assume a teacher in a manufacturing enterprise course organizes students into small groups to help select a product to be

**STAGE 1: Engagement**
Time during which students acquire information and engage in experiences that provide the basis for, or content of, their ensuing learning. Input may be provided in the form of teacher lectures, reading, video, demonstrations, field trips, etc.

**STAGE 2: Exploration**
Here, students have the opportunity to make an initial exploration of the new information, making tentative judgments as they bring their past experience and understandings to bear.

**STAGE 3: Transformation**
Learners continue to work in small groups, with the teacher asking the students to use and understand identified information. Students focus their attention on the aspect of the information which leads to the desired outcomes.

**STAGE 4: Presentation**
Learners are asked to present their findings to an interested and critical audience. Students explain what they have learned during the process. Explaining what they have learned to an interested group reinforces their understandings. In addition, student feedback enables them to evaluate their work.

**STAGE 5: Reflection**
By looking back at what they have learned and examining the process, students gain a deeper understanding of both the content and the learning process itself. This should help them with future learning.

*Figure 7-8: A five-stage model for cooperative learning.*

manufactured. Students would become product development teams and also serve as members of the final review team. This teaching/learning situation can be analyzed by applying the five-stage cooperative learning model as shown below:

Stage 1: Engagement
Teacher leads a discussion on the characteristics of a successful product. A design brief may be presented that provides material and cost limitations.

Stage 2: Exploration
Students establish guidelines for how the group will function. The group reflects upon their own interests and how their talents can be incorporated into the product development task. Multiple ideas are explored and the optimal suggestions are incorporated into a potential design.

Stage 3: Transformation
The group will select the specific product to be manufactured that relates to the group's interests and design criteria.

Stage 4: Presentation
The group presents their product suggestions to the entire class for approval. This step requires each group to prepare a rationale of their individual ideas and final product suggestion.

Stage 5: Reflection
The group might discuss the following questions: What did you learn about the characteristics of an effective product? What did you learn about the process of selecting a product to be manufactured? Was it helpful to receive input from other individuals in the group? What plan of action would you take if you had to repeat the process?

Cooperative learning is an appropriate teaching/learning strategy for studying manufacturing technology and its basic tasks—design, material processing, management activities, etc. Cooperative learning should be considered as an effective strategy for developing higher order learning skills and in developing collaborative skills in students. Young learners can also benefit from greater personal growth in acceptance of self and others. Perhaps more importantly, cooperative assignments allow students to develop crucial leadership and followership skills.

## Interdisciplinary Approach

One of the major challenges facing teachers is the formal integration of selected disciplines into existing manufacturing programs. Zuga (1988) noted that "relating concepts from other subject matter areas to technology education is called the interdisciplinary approach" (p. 56). A manufacturing technology program has the potential to integrate a wide variety of subject areas. A review of recent professional journals and conference proceedings indicate an emphasis upon the interdisciplinary relationship of math, science, and technology. In addition, national reports outside the field of technology education have specifically addressed the relationship between the disciplines of math, science, and technology. The American Association for the Advancement of Science published a report entitled, *Project 2061: Science for all Americans* (1989). It is a point of interest that the preface of the publication includes this statement:

> The terms and circumstances of human existence can be expected to change radically during the next human lifespan. Science, mathematics, and technology will be the center of that change—causing it, shaping it, responding to it. Therefore, they will be essential to the education of today's children for tomorrow's world. (p. 1)

*Rationale for an Integrated Curricula.* The National Council of Teachers of Mathematics has published *Curriculum and Evaluation Standards for School Mathematics* (1989) and *Professional Standards for Teaching Mathematics* (1991). The first publication addresses the standards for mathematics from grades K through 12. The Council speaks to a broad curriculum which offers experiences with different branches of mathematics included at all levels. Knowledge is linked to a conceptual framework which allows the use of problem-solving activities. The report also encourages that connections be made between various branches of mathematics and other curriculum areas. The second document encourages math teachers to create a dynamic environment that will make the subject matter come alive for every student. This includes cooperative learning; using logic, reasoning, conjecturing, inventing, and problem solving; and connecting mathematics and its applications instead of teaching isolated concepts. The emphasis of these two reports towards cooperative learning and problem solving makes a solid rationale for the application of mathematics in the study of manufacturing technologies.

The integration of math, science, and technology is linked to three important assumptions that possess a high degree of relevance to technology education teachers and professionals at all levels. First, there is a strong linkage between the study of modern technology and the appropriate application of math and science concepts. Second, technology education with an activity-based methodology, can provide an excellent integration of math, science and technology content into a relevant, meaningful, and functional learning experience. Third, the current emphasis on mathematics and science provides a unique opportunity for technology education to establish itself as a viable discipline to be studied by all students.

It is important to note that the interdisciplinary approach goes beyond the idea of producing consumer or industrial products. There are some technology teachers who could argue that they have always integrated math and science principles into their learning activities. However, most of these teachers have not effectively identified and communicated the specific interdisciplinary principles taught. Teachers and students need to reach beyond the specific product design or processing activities to identify and describe relevant math and science principles relative to the manufacturing activity.

The interdisciplinary approach also provides manufacturing educators with a unique opportunity to establish a closer partnership with the total school curriculum. The difficulty of addressing any technological development within a single discipline makes an effective argument for cooperation between programs and faculty. The importance of cooperation between disciplines was emphasized by Loepp (1991) in his article entitled, "Time for Action: Mathematics, Science and Technology," when he stated: "It seems to me that we have allowed the rather firm walls between disciplines to hinder the connection of knowledge to be made by learners" (p. 7).

*Approaches to Curriculum Integration.* The integration of school curricula to enhance learning is not new to the field of education. Sigurdson (1962) reported that the University of Chicago Laboratory school showed interest in the unification of mathematics and science at the turn of the century. More recently, movements such as writing and reading across the curriculum have become common practice in educational circles.

The integration of subject matter is organized into five distinct approaches (Dossey, 1991). In the simultaneous model, students simultaneously take courses in different disciplines and teachers deliberately make ties between the content of the courses. The braided model views content to be individual strands which are visited each year on a rotating pattern to develop a spirally-organized curriculum. In the topical model, the curriculum focuses on various topics throughout the year and all disciplines contribute to the development of a full understanding and application of the topic. The unified model for curriculum integration involves teachers from two or more disciplines to work together to identify a set of unifying ideas. This method often uses team-teaching techniques to present the subject matter. The fifth approach is the interdisciplinary model, which totally merges the subject matter from two or more disciplines. Using this approach, significant blocks of the school day are devoted to interdisciplinary studies, rather than specific time allocations for distinct classes in science, mathematics, technology, etc.

In technology education, numerous curriculum guides have encouraged the interdisciplinary approach. However, in most cases technology educators have not been successful in identifying and presenting interdisciplinary education. One exception, Dr. Donald Maley, has provided leadership for the development of teams of mathematics, science, and technology teachers through the development of instructional curricula. One example is the publication, *Math/Science/Technology Projects for the Technology Teacher* (1985). The booklet provides examples of interdisciplinary units that are appropriate for technology education.

*Developing an Integrated Instructional Team.* One of the most effective strategies for implementing an integrated manufacturing curriculum is through the use of an instructional team. The use of a team is appropriate for both the unified and interdisciplinary models presented in the previous section. This strategy is advantageous for several reasons. First, most manufacturing teachers have not been adequately prepared to provide instruction in the disciplines of either math or science. Secondly, the teacher does not realistically have the time to undertake such a massive curriculum change. Finally, the success of any integrated approach depends heavily on the coordination of the various disciplines toward a common area of study. The instructional team should adequately address the coordination issues and then focus on team goals.

The process of developing an instructional team to implement an integrated math/science/technology curriculum can be organized into at least five phases. The suggested sequence of activities are illustrated in Figure 7-9.

In summary, recent developments within the areas of mathematics, science, and technology education offer strong support for an integrated manufacturing curriculum. Math professions are calling for an increase in the applications-oriented approach. At the same time, science educators are encouraging an experientially-oriented curriculum with an emphasis on hands-on learning. Finally, manufacturing educators should recognize the value of strengthening their programs by including the application of mathematics and scientific principles in their classroom and laboratory activities. Through the integration of mathematics, science, and other disciplines, the manufacturing curriculum can become more meaningful and relevant to students.

| | |
|---|---|
| **PHASE ONE** | Involves the selection of individuals to serve on the team. Ideally, one teacher from each of the disciplines should be chosen. Individuals selected should possess an adequate knowledge of their discipline and an ability to work effectively in groups. It is fair to assume that the integrated curriculum effort will only be as good as the members of the team. |
| **PHASE TWO** | Involves the identification of the content and/or concepts appropriate for study. The manufacturing education teacher should assume the major responsibility in this process. Potential areas include: manufacturing systems, materials, processes, management, and the enterprise. |
| **PHASE THREE** | After the subject material and activities have been selected, teachers should identify and explain the respective mathematic, technological, and scientific principles. Maley (1985) suggests that this information be organized into information sheets that include the formulas and graphic examples of the principles. |
| **PHASE FOUR** | In this phase, the content and methodology of the learning units is checked. Each integrated unit should be pilot-tested with students. Members of the instructional team should not be the teachers of the pilot program. This makes certain that the comments and criticisms are unbiased. |
| **PHASE FIVE** | Involves the revisions of pilot-tested learning units and dissemination of the materials. The learning units should be disseminated to curriculum coordinators and department heads of the three disciplines involved. |

*Figure 7-9: Suggested process for developing an integrated math/science/technology instructional team.*

## Problem Solving

During the past decade, problem solving has become a major focus of the manufacturing technology curriculum. In this era of exponential growth, it is more important than ever for students to learn how to approach and develop solutions to technological problems. To develop an understanding of problem solving, it is helpful to review four major components including:

a) what is problem solving?, b) what is a problem?, c) what makes a good problem?, and d) the process of problem solving.

*What is Problem Solving?* Krulik (1987) describes problem solving as a process. It is the means by which an individual uses previously acquired knowledge, skills, and understanding to satisfy the demands of an unfamiliar situation. In a school environment, the student must synthesize what he or she has learned, then apply it to the new and different situation. Therefore, problem solving becomes an effective instructional strategy for the teaching of higher-order thinking skills.

Some educators assume that expertise in problem solving develops accidentally after one solves a variety of problems. While this may be true in part, problem-solving skills should be considered as a distinct, human endeavor and should be taught as such. Knowledge of the components of a problem-solving model and successful practice using the model is important to solving complex problems. The knowledge related to manufacturing technology can be divided into several parts including a) information and facts and b) the ability to use information and facts to solve problems of a technological nature. This ability to apply information and facts is an essential part of the problem-solving process. Ultimately, problem solving requires analysis and synthesis. Furthermore, to succeed in problem solving is to "learn how to learn."

*What is a Problem?* A major difficulty in discussing problem solving as it relates to the area of manufacturing technology seems to be a lack of any solid agreement as to what constitutes "a problem." Typically, a problem is a situation that confronts an individual or group that requires a solution and for which the individual/group sees no apparent or obvious mean or path to obtaining the solution. The key phrase in this description is "no apparent or obvious . . . path." As most students advance through a technology education program, what were problems at an early stage may become exercises and are eventually reduced to questions.

Krulik (1987) described these three commonly used terms as follows:

- Question: A situation that can be resolved by mere recall.
- Exercise: A situation that involves drill and practice to reinforce a previously learned skill.
- Problem: A situation that requires thought and synthesis of previously learned knowledge to resolve.

In addition, Krulik indicated the problem must be perceived as such by a student, regardless of the reason or situation. A problem must satisfy the following criteria:

- Acceptance: The individual accepts the problem. There is a personal involvement which may be due to a variety of reasons including internal or external motivation, or simply the desire to experience the enjoyment of solving a problem.
- Blockage: The individual's initial attempt at solution is fruitless. His or her habitual responses and patterns of attack do not work.
- Exploration: The personal involvement as identified in the acceptance description forces the individual to explore new methods of attack.

*What is a Good Problem?* It should be apparent at this point that problems are the basic medium for problem solving. Furthermore, problem solving should be a basic skill of a manufacturing technology program. It follows then that without "good" problems we could not have an effective program.

First of all, teachers should be aware that good problems can be found in every area of modern manufacturing. Problems may also relate to various types and levels of learning. The four items listed below help outline some characteristics of good manufacturing problems. It should be noted that not every good problem needs to have all of these characteristics. Neither is it always possible to clearly identify which characteristics make a problem an appropriate assignment for a problem-solving activity. However, a good problem will always have these attributes:

- The solution to the problem involves the understanding of a distinct manufacturing concept or the use of a "hands-on" skill.
- The problem can be generalized or extended to a variety of situations.
- The problem lends itself to a variety of solutions.
- The solution to the problem requires the application of appropriate interdisciplinary concepts.

*Process of Problem Solving.* Problem solving has previously been described as a process that starts when the initial encounter with a problem is made and ends when the obtained solution is identified, evaluated, and successfully implemented. The importance of teaching this process was emphasized by Maley (1991) when he stated:

> The processes involved in problem solving are essential elements in a well designed experience in this dimension of Technology Education. The processes by which the students achieve the results or answers may be more important than the answers one gets. (p. 31)

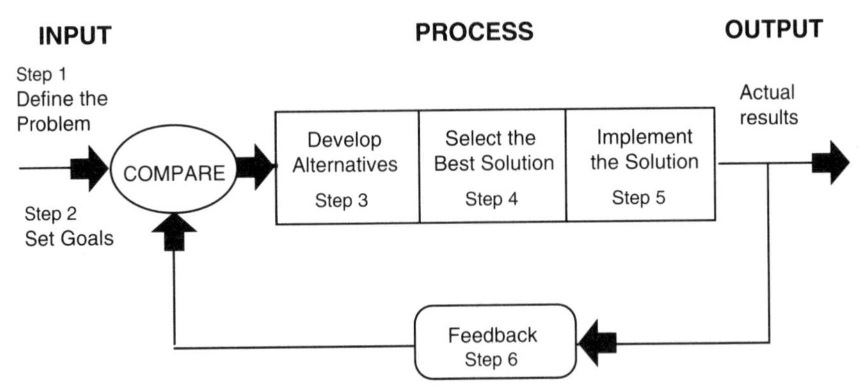


*Figure 7-10: Applicable problem-solving model for the study of manufacturing technology.*

The process of problem solving is complex and difficult to learn. It consists of a series of tasks and mental processes that are loosely linked together to form what is called a set of heuristics or a heuristic pattern. The process involves a set of guidelines that a student must follow in order to solve a problem. Although numerous heuristic patterns have been proposed, the model presented by Hacker (1987) in Figure 7-10 is appropriate for manufacturing programs.

Finally, it is important to identify those characteristics that make an effective teacher of problem solving. Students involved in complex problems often depend on the teacher for encouragement and continual support. Without an interested, energetic, and involved teacher, nothing constructive may take place. Problems should be solved in the laboratory with the teacher allowing for and encouraging a wide variety of approaches, ideas, questions, solutions, and discussions.

Success in problem solving requires a teacher with a positive attitude towards problem solving itself. Some teachers dislike problem solving because they have not had enough successful experiences in this area. Teachers must prepare carefully for problem solving. Manufacturing teachers who encourage their students to solve problems, who make students think, and who ask open-ended questions will provide their students with valuable problem-solving experiences.

## STRATEGIES FOR EVALUATION

While teaching manufacturing, evaluation data may be obtained by presenting an individual with a set of given tasks to perform, by asking questions about themselves, or by asking others to observe and judge the behavior. These three methods of obtaining data establish the three most appropriate types of evaluation strategies. The three strategies are commonly referred to as testing procedures, self-report techniques, and observational techniques, respectfully.

Testing procedures are designed to assess cognitive learning. According to Gronlund (1969), "a test is merely a series of tasks which is used to measure a sample of a person's behavior at a given time" (p. 11). The most commonly used tests in a manufacturing technology program are teacher-prepared, written, achievement tests. When written properly, these tests are able to accurately assess cognitive subject matter presented.

Self-report techniques allow the teacher to gain valuable information about an individual or group. For evaluation purposes this information is usually obtained by interview or questionnaire. The interview involves a face-to-face relationship between the teacher and the student(s). Although this method can be extremely effective for gathering information quickly, it is usually inappropriate for large group situations. The questionnaire method of obtaining information is most commonly used in systematic attempts to assess interests, attitudes, and other aspects related to the teaching/learning process. It has a distinct advantage over the interview in that it is effective for most group situations. It should also be noted that since both of these techniques depend upon the willingness of the individual(s) to provide honest answers, they are somewhat limited when evaluating student progress and are most effective for program and instructional assessment activities.

Observational techniques provide an effective means for gathering reliable data about student behavior. Observation techniques are merely systematic methods of recording observations for purposes of evaluation. Anecdotal records, checklists, rating scales, and sociometric techniques are included in this category. Although most observational techniques are performed by the teacher, it is often appropriate to obtain peer evaluations. Since the manufacturing program often involves cooperative learning in small groups, peer evaluation in addition to teacher's observations are meaningful. The chart in Figure 7-11 illustrates how a teacher-made peer evaluation response form can be used to assess individuals' contributions to a cooperative learning experience.

In conclusion, educational evaluation plays an important role in the manufacturing technology program and it provides information which serves as a basis for a variety of educational decisions. The main emphasis in the evaluation process is the student and his/her learning progress.

**RATING SCALE:** 5 always
4 frequently
3 sometimes
2 seldom
1 never

| Student Name | | | | | |
|---|---|---|---|---|---|
| Attends class on time and makes effective use of time | | | | | |
| Effectively performs duties and responsibilities assigned by the group | | | | | |
| Makes effective use of time outside of scheduled lab periods | | | | | |
| Makes effective use of scheduled work periods | | | | | |
| Effectively communicates with fellow group members | | | | | |
| Willingness to assume leadership roles | | | | | |
| Willingness to assume followership roles | | | | | |
| Contributes to development of the construction packet | | | | | |
| Solves technical problems | | | | | |
| Procures needed materials and information | | | | | |
| Overall contribution to the success of the product | | | | | |

*Figure 7-11: Sample form for evaluating cooperative learning experiences.*

# SUMMARY

An effective manufacturing teacher needs to employ a wide range of instructional strategies in order to meet the needs of today's students. The three major factors that can be used to assess the needs of students include: attention sets, motivation, and developmental tasks.

The process of developing an instructional strategy can often be made easier if accomplished in a systematic fashion. A model for planning instructional strategies for a manufacturing technology program involves six basic steps: a) selecting content, b) identifying the type of learning, c) writing an objective, d) selecting an instructional strategy, e) organizing content, and f) evaluating and assessing the teaching/learning process.

Traditional teacher-directed instructional strategies such as the lecture and demonstration need to be supplemented in order to encourage students to develop higher-order thinking skills. Suggested instructional strategies that are appropriate for enhancing thinking skills in manufacturing technology programs include conceptual learning, cooperative learning, problem-solving techniques, and applying an interdisciplinary approach.

# REFERENCES

Bloom, B. S. (1956). *Taxonomy of educational objectives*. New York: David McKay Company, Inc.

*Curriculum and evaluation standards for school mathematics.* (1989). Reston VA: National Council of Teachers of Mathematics.

Dossey, J. A. (1991). *Mathematics and science education: Convergence or divergence*. Unpublished document.

Ebel, Robert L. (1982). *Essentials of educational measurement*. Englewood Cliffs, NJ: Prentice Hall, Inc.

Gagne, R. M. (1970). *The conditions of learning* (2nd ed.). New York: Holt, Rinehart and Winston, Inc.

Gronlund, N. E. (1965). *Measurement and evaluation in teaching*. London: Collier-MacMillan Limited.

Gunther, M. A., Estes, T. H., & Schwab, J. H. (1990). *Instruction: A models approach*. Boston, MA, Allyn and Bacon.

Hacker, H., & Barden, R. (1989). *Technology in your world*. Albany, NY: Delmar Publishers.

Havinghurst, R. J. (1974). Developmental Tasks. In G. Hass, J. Bondi, & K. Wiles, (Eds.) *Curriculum planning: A new approach*. (pp. 54-72). New York: Allyn and Bacon, Inc.

Henak, R. (1988). Cooperative group interaction techniques. In W. Kemp & A. Schwaller (Eds.) *Instructional strategies for technology education*. (pp. 143-165). Council on Technology Education Yearbook, Mission Hills, CA: Glencoe Publishers.

Israel, E. N. (1972). An examination of nonverbal and verbal instructional guidelines in students developing abstract understanding of technical concepts. *Dissertation Abstracts International, 33.* 3343-A—3344-A (University Microfilms No. 73-823).

Johnson, D., Johnson, R., Johnson, E., & Roy, P. (1984). *Circles of learning: cooperation in the classroom*. Alexandria, VA: Association for Supervision and Curriculum Development.

Klausmeier, H. J. (1976). Instructional design and the teaching of concepts. In J. R. Levin and V. L. Allen (eds.). *Cognitive learning in children.* New York: Academic Press.

Krulik S., & Rudnick, J. (1987). *Problem solving: A handbook for teachers*. Boston, MA: Allyn and Bacon, Inc.

Loepp, F. (1991). Time for action: Mathematics, science and technology education at Illinois State University. *The Technology Teacher*, *51*(2), 5-8.

Mager, R. F. (1962). *Preparing instructional objectives*. Palo Alto, CA: Fearon Publishers.

Maley, D. (1991). *Value in technology education*. Baltimore, MD: Maryland Vocational Curriculum Research and Development Center.

Maley, D. (1985). *Math/science/technology projects for the technology teacher*. Reston, VA: International Technology Education Association.

Maley, D. (1985). Interfacing technology education, mathematics, and science. *The Technology Teacher, 44*(11), 7-10.

Maley, D. (1976). *Student development: an essential base for program development*. Unpublished paper presented at American Industrial Arts Association Conference, Des Moines, IA.

*Professional standards for teaching mathematics.* (1991). Reston, VA: National Council of Teachers of Mathematics.*Project 2061: Science for all Americans*. (1989). Washington, D.C: American Association for the Advancement of Science.

*Project 2061: Science for all Americans.* (1989). Washington, D.C.: American Association for the Advancement of Sciences.

Raths, L. E., Jonas, A., Rothstein, A. & Wassermann, S. (1967) *Teaching for thinking, theory and application*. Columbus, OH: Charles E. Merrill.

Reid, J., Forrestal, P., & Cook, J. (1989). *Small group learning in the classroom*. Concord, Ontario: Irwin Publishing.

Sigurdson, S. E. (1962). *The development of the idea of unified mathematics in the secondary school curriculum 1890-1930.* (Doctoral dissertation, University of Wisconsin-Madison).

Silver, H. F., & Hanson, R. (1991). *Teaching styles and strategies*. Moorestown, NJ: The Institute for Cognitive and Behavioral Studies in association with Hanson, Silver and Associates.

Slavin, R. (1980). *Cooperative learning: What research says to the teacher*. Baltimore, MD: Center for Social Organization of Schools.

Slavin, R. (1987, March). Cooperative learning: Can students help students learn? *Instructor,* pp. 74-78.

Wright, R. T. (1991). *Manufacturing systems.* South Holland, IL: Goodheart-Willcox.

Zuga, K. (1988). Interdisciplinary Approach. In W. Kemp & A. Schwaller (Eds.), *Instructional strategies for technology education*. (pp 56-71). 37th Yearbook, Council on Technology Education Yearbook, Mission Hills, CA: Glencoe Publishing.

*CHAPTER* **8**

# *Facilities For Teaching Manufacturing*

Dr. Douglas Polette (Professor)
*Montana State University, Bozeman, MT*

By its very nature, technology education requires the learner be actively involved in their learning. Consequently, there is a need to have quality, well-planned facilities at all levels of education. Although the specifics will vary from the elementary through collegiate program, a common thread runs though all manufacturing technology facility designs. The facility should be a place where students can learn about the dynamics of manufacturing; that is, how ideas are converted into useful products in a managed system. Along this educational journey, the students may use computers to design products, construct prototypes, evaluate specific materials with sophisticated testing apparatus, plan and build jigs and fixtures, prepare a quality assurance program, set up and operate a production line, and watch instructional media. A facility to encompass this variety of activities should be developed only after careful planning and implementation efforts.

This chapter will address the importance of the physical environment necessary to support the instructional activities suggested for each level. Particular emphasis will be placed on a typical middle school and high school facility since these areas often present the greatest challenges. Suggestions will be offered for renovating older facilities as well as designing and equipping new manufacturing areas.

## Facilities Support an Approved Curriculum

Instruction of manufacturing topics should occur in a facility that is designed for the single, basic purpose of addressing manufacturing technologies. The laboratory should be a major component in a unified technology education program. Instructional efforts in the manufacturing area should compliment the department's approved curriculum. Today, more than ever, the public will not (and should not) support facility improvements without a sound justification of the school's instructional programs. While this chapter will deal with the actual design of the facilities, it is assumed that the reader first has a complete understanding of the philosophical basis established in the preceding chapters. Without this foundation, a quality manufacturing program cannot be established.

A careful analysis of the curriculum must be completed before improving facilities. The curriculum helps determine the physical requirements of the manufacturing area and should address those issues which link the planned activities to be conducted in the facility and the educational goals of the program. A well-defined curriculum will address the needs of the students and community. Regardless of the level, it is imperative the manufacturing facility be designed to meet the needs of the students.

Since educational systems are not able to keep pace with the dollars spent on equipment and facilities in the manufacturing sector, it is critical that school facilities be developed to support a number of diversified learning experiences. They must be comprehensive, flexible, and capable of supporting both present and future needs. A careful analysis and integration of available resources will allow for the best value for the dollar. But, it must be emphasized again that a well-defined curriculum should be specified before developing plans for new or remodeled facilities, Figure 8-1.

# CHARACTERISTICS OF THE MANUFACTURING FACILITY

To best utilize the space reserved for a technology education program, the available space should be subdivided into areas related to specific instructional tasks and priorities. These areas, of course, must be based on the arrangement of the curriculum for a particular program or educational level. Currently, the most common arrangement of curriculum centers around the four cluster areas of communication, construction, manufacturing, and transportation. Additional areas at the collegiate level might include design laboratories for research and development efforts and a room for professional (teacher preparation) experiences. Another arrangement that is gaining favor is to develop dedicated laboratories around themes such as physical, bio-related, and information technologies. No matter what curriculum arrangement, there is a need for the student to develop an understanding of manufacturing systems and their impact on the environment, economy, and society.

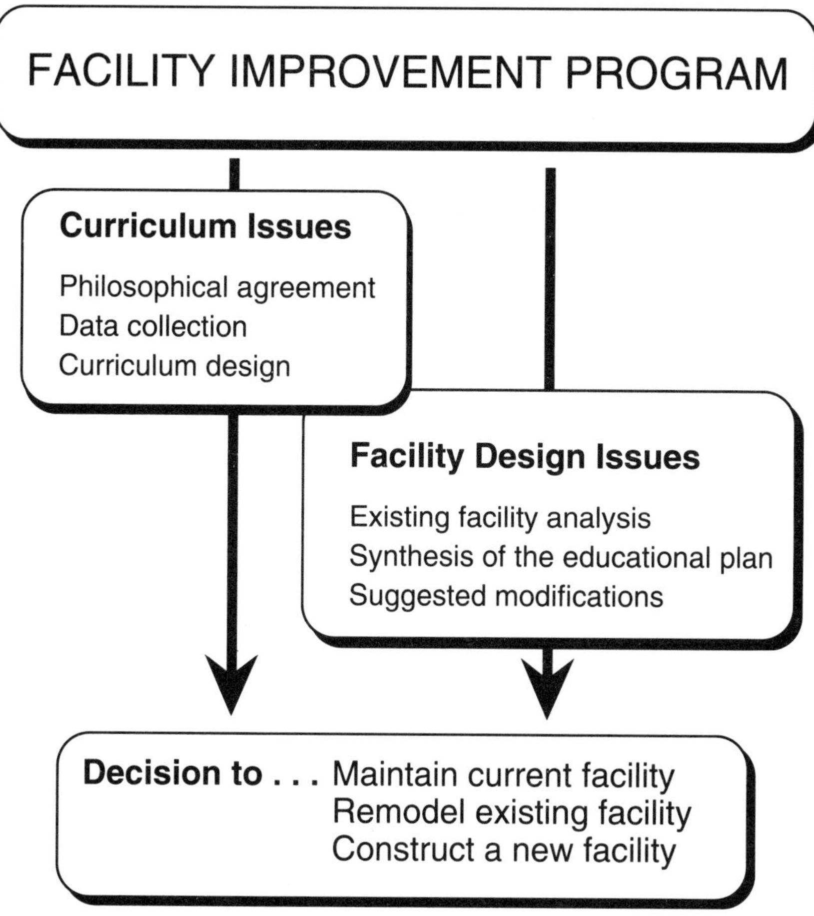


*Figure 8-1: Issues and required decisions in a facility improvement program.*

To support the approved curriculum, the facility should include more than just the manufacturing laboratory itself. Ancillary areas must be carefully designed as part of the total manufacturing facility. These auxiliary areas generally include the resource area, a classroom, the material processing laboratory, material and product storage areas, and the office area.

## Design and Resource Area

Activities conducted during the design phase of manufacturing courses require the use of multiple resources such as reference books, videotapes, computers, simulation software, modems, etc. Various resources and media are best housed in a separate room that can be closed off from the main laboratory to limit the infiltration of dust and noise. At the collegiate and secondary level, this area could be incorporated into a classroom adjacent to the manufacturing lab. At the middle school level, the design and resource area may become part of a combined communication/manufacturing section of the technology education department. At the elementary level, design activities are routinely conducted in a section of the classroom.

Generally, it is desirable to separate the major noise and dust-producing area(s) from the rest of the manufacturing facility with the use of glass partitions which allow visual control but restrict the flow of dust and noise. This is especially important since design tasks involve materials and hardware that should be kept clean—computers and printers, diskettes, reference books, engineering drawings, etc. Elementary design activities are generally such that they fail to produce the quantity of dust and noise often found at higher levels.

The manufacturing design and resource center might evolve from a traditional drawing room since the drafting tables are useful for the preparation of graphic media. If computer workstations are also available in the area, they may be used for creative tasks and documenting product and system designs. This room typically serves as a store room for media, extra software, and audio-visual equipment.

## Formal Instructional Area

Formal instructional areas generally include the classroom where standard instructional functions (taking attendance, lectures, etc.) are carried out. The classroom should be large enough to accommodate up to 24 students. The furniture preferred is movable tables and chairs. This type of furniture provides the most flexibility for formal lectures, small work groups, or total class activities. Tables provide a large, flat surface area for working on design problems and can also hold computers, portable drawing boards, and similar equipment as needed. It is desirable to have the classroom directly adjacent to the laboratory with glass partitions separating the two for visual control by the instructor, Figure 8-2.

In many cases, discussions, seminars, and small group activities can be conducted in this area. While many of these tasks can be incorporated into other facility areas, it is useful to reserve a separate section in the manufacturing area for these purposes. The furniture in this area should include flexible chairs and tables to allow the furnishings to be changed to match individual and group demands.

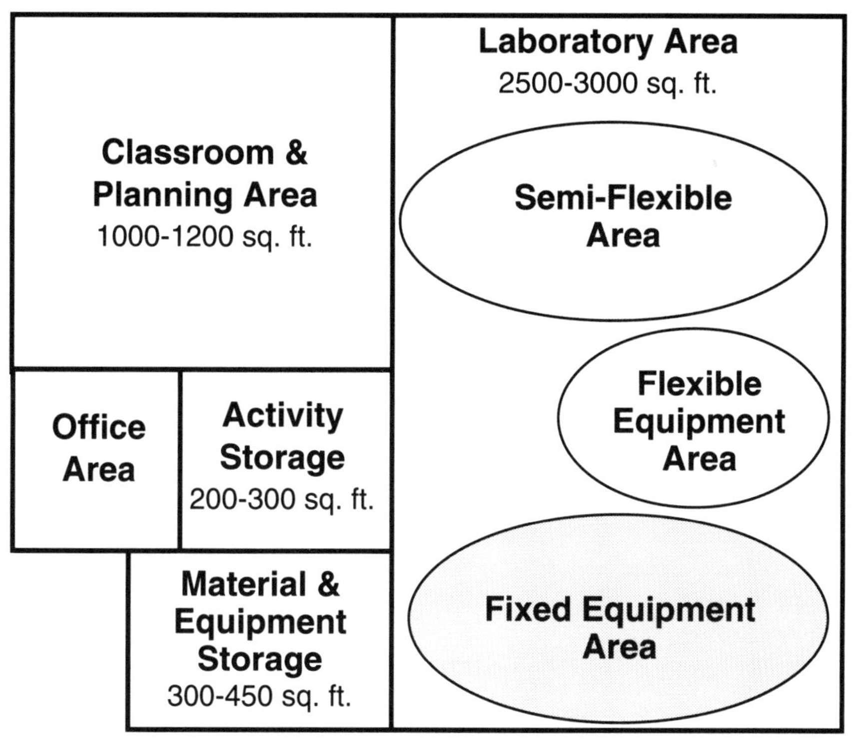


*Figure 8-2: Recommended areas and space allocations for a secondary education manufacturing facility.*

At the elementary level, the regular classroom often serves as both the formal and informal instructional area as well as the manufacturing laboratory. This is not a hindrance since younger students are used to a variety of experiences in their home classroom (e.g., music lessons, art assignments, etc.). Most manufacturing experiences at this level involve hand tools and tabletop systems so dedicated laboratories are not always necessary.

## Material and Product Storage Area

Storage rooms are generally the areas of either new or remodeled construction projects that are eliminated first when cost-cutting measures begin. This is unfortunate since it forces the use of the laboratory for storage of materials and student work. Instructional materials stored in a main traffic area are difficult to control and maintain. The same problem exists when storing student and class products in a prime location.

The amount of material storage space can best be determined as a ratio to the size of the main lab. It is recommended that up to 20 percent of the manufacturing facility be dedicated to the storage of materials, supplies, and student work. The storage area should be long enough to accommodate the full lengths of various materials. Sheet materials are best stored flat but that requires additional space. To economize on the storage of plywood and sheet metal, vertical storage racks may be specified. A series of lockable cabinets are also desirable to store specialized and more expensive items. Also, if flammable liquids and paints are to be safely stored in the facility a specially designed cabinet needs to be part of the storage area. Compressed gases will also need a suitable area where they can be secured with a proper distance between the bottles.

The volume of student work fluctuates considerably in a manufacturing program. For instance, an enterprise class may produce 100 products in a single class period, so the demand for storage space may increase suddenly. Then as members of the enterprise distribute their completed products, storage problems disappear just as quickly. In other coursework, students may require flexible space for individual or small group work. The listing below indicates several of the storage problems that might arise in the manufacturing program:

- Storage of numerous jigs and fixtures from a systems design class,
- Space for developed marketing materials (posters, flyers, etc.),
- Storage of pre-printed packages ready for final assembly/packaging tasks,
- Space to maintain a product inventory from a production run,
- Shelves or lockers for individual material testing samples,
- Storage of paperwork from enterprise and production planning efforts, and
- Space to store defective parts and work-in-progress from production runs.

## Material Processing Area

Material processing, finishing, assembly, and packaging are often best accommodated within the main laboratory. Because these activities should be incorporated within production activities, it is desirable if they can be moved to their appropriate location during student work sessions. The main laboratory area needs to have storage areas for the equipment within the laboratory plus an area where surplus and seldom-used equipment can be stored out of the way until needed. An efficient manufacturing program requires flexibility within the laboratory so it can be easily rearranged in a relatively short period of time to accommodate the varying needs of the space.

The manufacturing laboratory is perhaps the most diverse of all the cluster laboratories. This is primarily due to the enormous variety of tools, equipment, and materials needed to support the variety of products students may select to manufacture. Furthermore, the equipment must be able to be moved to different locations in order to promote efficient flow of materials (similar to the production line concepts common in industry). Thus, the laboratory should have flexible components with the ability to connect the equipment and tools to the required utilities at various locations within the room.

A standard manufacturing facility will include three general types of equipment, Figure 8-2. First is the permanently positioned (fixed) equipment that must remain in one place due to size, weight, or utility connections. This category includes items like milling machines and ceramic kilns. Second is the flexible but larger (semi-rigid) devices and equipment. Although awkward, most benches and larger equipment can be moved to a desired location as needed to support manufacturing activities. The third grouping includes all flexible equipment that are easily positioned for production activities. Common items in this third category include conveyor belts, packaging machines, and any equipment mounted on wheels.

The specific size of the main laboratory will depend on the age level of the students and how the total technology education program is structured. At the elementary level the most common approach is to conduct laboratory activities within the classroom itself unless a separate facility is provided for the class activities. At the middle-school level, the general technology laboratory "becomes" the manufacturing facility when manufacturing activities are conducted. In this case, the laboratory should have a minimum of 125 square feet for each student exclusive of ancillary space. The same figure is appropriate at the high school level and should be increased for a collegiate manufacturing program.

The generally accepted square footage requirement of a manufacturing laboratory is from 90 to 150 square feet (per student) with 125 square feet considered by most adequate to provide enough space to perform normal manufacturing activities. Also, additional areas should be available for the storage of hand tools, specialized equipment, etc. The optimum space for a class of 24 students would be up to 3000 square feet plus another 500-800 square feet for storage. The shape of the facility might be in a rectangle with a length/width ratio of approximately 2 to 3. This provides for the best and most efficient arrangement of equipment and benches for instruction and supervisory purposes. Any shape that does not allow the instructor to have full visual control of the students should be avoided.

## Office Area

Modern manufacturing requires extensive use of information technology and the sharing of data with other firms and businesses. It is no different in the typical manufacturing laboratory. At the elementary level, a phone should be available in the classroom so the teacher can communicate with others about class activities. At the middle school level, where the instructor is involved in the ordering of tools and materials, it is important for the instructor to have an office area where phone conversations and computer networking can be carried out without the distractions of classroom and equipment noise. At the secondary and university level, the instructor should have an enclosed office for the above reasons as well as the privacy required when working with students.

The office area also must serve the instructional pursuits of the manufacturing teacher. Therefore, sufficient space should be available to storage and maintain audio-visuals, reference books, supply catalogs, and miscellaneous examples of prior student work. In this era of professional and service activities, a computer and printer is essential. Also, a drawing board, photocopier, and light table are convenient.

## Additional Characteristics of Manufacturing Facilities

Besides the classroom and laboratory areas dedicated to teacher/student activities, there are several additional characteristics of the typical manufacturing facility. This list includes necessary utilities (electricity, compressed gases, etc.) and the climate control systems to regulate dust, humidity, and fumes. This section also addresses issues such as safety, accessibility, security, and noise control.

Utilities available in a manufacturing laboratory help transform the available space into a learning environment. The specific utilities needed will typically depend on the specific activities that will take place in the classroom. However, care should be exercised to assure that utilities installed at the time of construction or remodeling will be adequate for future program requirements. A future-oriented facility will not only include the usual utilities such as phone, electricity, water, and waste systems but may also include communication links, specialized safety systems, and a variety of instructional technologies.

In general, utility systems should be designed in such a way that any of the systems would be adequate for usage ten to twenty years in the future. Furthermore, the utility system should be installed in such a manner that maintenance and upgrading can be easily accomplished and without undue costs. Naturally, all systems must be designed and installed in accordance with the latest building codes. The following sections describe the major considerations of the vital utilities and their relationship to modern manufacturing laboratory facilities.

*Electricity.* Electrical system requirements include 110/120, 208, and/or 220 VAC with amperage ratings sufficient to supply adequate current to the equipment being operated. Both single and three phase power should be made available. Dedicated circuits may be required to operate specialized computer equipment where voltage surges may cause problems. Electrical circuits should be designed to provide adequate current when several pieces of equipment are operated simultaneously.

The location of receptacles should provide for maximum flexibility so equipment can be easily moved to accommodate different manufacturing set-ups. Also 110/120 VAC receptacles should be installed immediately outside of the facility to allow for those outdoor activities that may require electrical power. In all cases, up-to-date electrical codes must be followed with special considerations for the use of GFI (ground fault interrupter) receptacles.

Electrical "kill" switches may be required by local codes in those rooms where potentially dangerous equipment is being used. In this case, it is important to locate these switches in several locations throughout the facility, but one should be located next to the exit so the instructor may cut off all power if required to leave the room while students are in the facility.

*Water.* An adequate supply of both hot and cold water is essential for proper personal hygiene as well as many manufacturing activities. Emergency showers should be provided if individuals must be quickly decontaminated from contact with hazardous materials. For some activities the water may have to be filtered, treated, or conditioned in order to meet the requirements of a particular activity or to properly function in a piece of equipment. Generally this type of equipment can be added at the point of use and therefore would not entail an extensive utility system.

Drinking fountains should be provided in or near the facility and should be of the refrigerator type to provide water at the optimal drinking temperature. In and around areas where there is a potential for water spillage, the floor area should be of a non-slip type to help reduce accidents.

*Compressed Air and Combustible Gases.* The major combustible gases that may potentially be used in a manufacturing facility include natural gas, propane, and acetylene. Some of these gases may be supplied under pressure in bottled form or piped directly into the lab from supply lines. In either case, strict adherence to codes must be followed as some gases are not compatible with certain materials and other gases. Safety regulations must also be followed for the storing of full or empty gas cylinders to prevent potentially dangerous conditions. An air compressor should be located nearby but not in the main laboratory area. All supply lines must be properly marked and identified so no mixing of gases can occur in the laboratory.

*Climate Control.* Climate control consists of providing a physical environment which is considered to be optimal for humans to function at the best of their ability without undue stress from extremes of temperature, humidity, and ventilation. An adequate supply of fresh air, filtered and heated/cooled is critical to enhance learning. Gentle air movement is essential to restore the proper amount of oxygen in the air and to remove carbon dioxide and other contaminates. Specific building codes should be carefully adhered to as they often specify the required amount of air exchanges per hour needed in various classrooms and laboratories. Controls should be located so they can be adjusted by the teacher or other authorized personnel to meet the particular needs of the area. Special measures must be considered when it is also important to exhaust large volumes of air in order to filter out particles such as smoke and dust to make it safe for breathing.

Because of the diversity of the activities (that may be carried on in a technology education facility) that produce pollutants associated with photographic darkrooms, finishing rooms, foundry metals, engine exhaust, and plastic processing, it is important to design the air handling equipment with the above potential problems in mind. In some cases, air can be filtered and returned to the classroom while in other cases the exhaust air must be completely vented to the outside. Heating, ventilation, and air-conditioning professionals must be consulted to assure the entire system functions efficiently and to proper standards. Specific industrial standards must be followed in accordance with existing building codes.

Additional measures should be taken to maintain proper moisture content to not only make it comfortable for students and teachers but also for the proper functioning and maintenance of equipment. In some climates, the humidity is high enough to rust unprotected metal surfaces, while in other climates the humidity is so low that it causes static electricity problems with computer equipment. Whatever the local conditions, some measures may need to be taken to assure a quality learning environment.

Although at present, educational institutions are not heavily used during the summer months, the year-round school is a real possibility. Therefore, heating, ventilation, and especially air conditioning equipment should be designed with this in mind.

*Noise Reduction.* Excessive noise can be both distracting and harmful to the occupants of a manufacturing facility. Noises not only distract from a quality learning environment, but also can be physically damaging to individuals. The best approach to dealing with equipment which produces excess noise is to prevent or reduce the amount of noise at its source. This may be accomplished by purchasing equipment with noise-absorbing features or modifying the existing environment. Sound-absorbing treatments on the walls, ceilings, and floors will help reduce the reflective noise and help to quiet the whole area. One effective technique is to locate noisy equipment in the facility away from areas that need to have a quieter environment. Few large, traditional projects are produced in manufacturing technology programs; consequently, the opportunity exists to purchase smaller portable or tabletop equipment which produce less noise. Noise standards established by OSHA and other agencies should be followed.

*Telephone.* The telephone in the manufacturing facility is a necessity in today's information society. It is desirable to have easy access to phone connections for both student and instructor use with adequate circuits to provide communication links within the classroom, between classrooms, and to the community. Telephone cables supply a means of voice communication plus also carry digital signals for communication between computers, modems, and local area networks (LANs). The local phone company should be contacted early in the planning phase to assure proper design and adequate lines for future expansion.

*Satellite, Video, and Interactive Video.* With the advent of modern telecommunications, it would be helpful if the technology laboratory is linked to a satellite receiving dish and/or local cable TV service. The video lines should run to the resource area and formal classroom. Since video recordings represent such an excellent teaching tool, it is also important to be able to send video signals to classrooms throughout the school from a studio located in the technology education department. Therefore, connecting cables need to be installed during construction or remodeling.

*Accessibility for Special Needs Students.* All students can profit from and contribute to a manufacturing program. Physical or mental barriers should be removed so the program can be available to all students. For instance, work surfaces and machinery should be installed so that wheelchair-bound students can easily reach important parts or features. Also, vises placed on the edge of tables should not hinder the flow of traffic around the facility. Tool panels and storage areas should be designed to account for different student needs.

The right of the special needs learner was originally assured by the Education for All Handicapped Children Act, PL 94-142 (1975). But more recently approved legislation has dramatically changed the nature of accessibility guidelines in public buildings. It is imperative that all local and federal statutes be followed during renovation or new construction.

In the design of any facility, forethought is required to not place artificial barriers which inhibit the accessibility of special needs individuals from full participation in the activities of the curriculum. A professional monograph "Making Industrial Education Facilities Accessible to the Physically Disabled" edited by Shackelford and Henak (1982), is available from the International Technology Education Association Reston, Virginia and should be consulted when planning new facilities.

*Safety Considerations.* Safety is one of the prime considerations in all school systems. Any manufacturing facility, regardless of the age of the student, should be safely organized and managed. Because the learner will be working with a variety of tools, materials, and products during their manufacturing experiences, there is a chance that accidents may happen. Safety glass cabinets, first aid kits, and related emergency materials should be placed in a visible and appropriate place. Also, the area must be easy to clean and attractive to the student, the instructor, and the general public.

By its very nature the manufacturing laboratory includes a large number of materials and utilities which if improperly stored, used, or mixed may result in the potential for fire. To further complicate the scene, many laboratories contain computer equipment or other sophisticated apparatus that will be severely damaged if certain types of fire-suppression devices are used. This situation then requires very specialized analysis of the type of fire extinguishing system to be used in each laboratory. Fire professionals should be consulted to aid in the design of any fire-suppression equipment or devices to assure that they will provide the protection needed if and when the need arises. Fire alarm boxes and extinguishers should be clearly marked and conveniently located in such locations where they will not be obscured by equipment. Naturally each room should have an evacuation plan plainly posted which provides directions for building exit and assembly points. In all states, school building plans for either new or remodeled facilities must be reviewed by personnel from the Fire Marshal's office to assure the building will meet all codes.

In most states, laws have been passed which require the inspection of the plans for any new or remodeled school facility. This is designed to assure the facility will meet specific safety standards established for the safety of the occupants of the school. These standards may vary from state to state but in general they will cover such items as fire exits, locations of fire-suppression equipment, the swing of doors, the handling of toxic substances and waste, ventilation and exhaust, etc. As plans are being developed, it is important to contact the proper state safety officials to assure the remodeled facility will meet all current safety standards.

*Security.* The security of the learner, the tools, equipment, and supplies need to be considered in specifying manufacturing facilities. This topic includes items as the keying of door locks, fences, specialized cabinets, and hardware locks on computer equipment; lighting and exit signs; and surveillance systems. Whatever special systems are installed, they must be designed so normal activities of instruction are not hindered in the classroom or laboratory. Security systems must also be designed to allow the instructor(s) to carry on instructional duties and activities which may take place during evening hours or on weekends.

*Additional Factors.* Wash basins and sinks should be located within the facility. Special filters or traps may be required in sinks where harmful chemicals, ceramic materials, or similar items are being used. Both liquid and powdered soaps should be available for students to clean up after manufacturing activities. In addition, rest rooms should be provided for both males and females in close vicinity to the laboratory and should meet all standards for accessibility for the handicapped. If adult classes might be offered, then the rest rooms area must not be closed during evening and/or weekend hours.

## PLANNING FACILITY IMPROVEMENTS

Before any physical changes take place within the manufacturing area, a great deal of time and effort must first be initiated to develop the background information needed to restructure the program. This process generally involves countless hours of discussions, writings, committee meetings, and presentations. Consequently, prior to making major adjustments in the facility, there should be some assumptions made among the "players" on the educational team. These assumptions are as follows:

- All faculty in the program arrive at consensus on the philosophical base for their technology education program.
- A curriculum plan has been reviewed and established for the manufacturing program within the total offerings of technology education in the school.
- The administration is in support of the proposed curriculum and facility changes or at least has an open mind to allow change to take place.

- The local educational agency has the funding base to support facility renovation, remodeling or new construction.

It is important to reiterate the relationship of the facility design to the curriculum. Though the philosophical base and the curriculum design may be well established long before the facility is designed, it is the facility that communicates to others what the curriculum is all about. Consequently, the laboratory has a major impact on how others view the program.

Technology education programs are multi-disciplinary in nature. Thus, facilities need to be capable of supporting a number of diversified learning experiences. To meet this challenge manufacturing facilities need to be comprehensive, flexible, and usable to support both present and future technology education curricula needs.

When designing and equipping facilities, facilities and equipment should be viewed as a planned resource that supports instruction and is an integral part of the instructional system. To reach this goal, it is suggested that teachers, administrators, and construction planners follow a set procedure to avoid missing key details. This section outlines a model for this design process, Figure 8-3.

## Determining Facility Needs

Determining the need for new facilities or the redesign and renovation of existing facilities is a complex process which must be made with the interests of the students in mind. Unfortunately, this process is often complicated if key people in the planning display a tendency to maintain many of the same areas, room arrangements, and organizational structure of which they are accustomed. Innovative solutions to existing facility problems are often best offered from individuals who have not been part of the program in the past.

Regardless of the type of manufacturing area, the needs of the entire program and manufacturing area must first be determined. Rokke (1988), in a paper delivered at a past AVA conference, discussed the appropriateness of conducting a needs assessment as one of the first steps in the design of a laboratory. He suggested the needs statement be used to identify the programs, course content, equipment, laboratory design criteria, and utilities required to support the proposed construction. Rokke compiled the following questions to initiate the facility design process:

1. What courses will be taught in the facility?
2. What types of equipment are required to teach the courses?
3. How many students will be taught?
4. How many teachers will be involved?

5. How much total space is required for each piece of equipment?
6. How much material and student storage space will be required?
7. What utilities are required?
8. What are the safety considerations?
9. What are the audiovisual, computer, and telecommunication requirements?
10. Will auxiliary rooms or areas be needed?
11. What provisions are necessary for climate control?
12. Will noise abatement be necessary?
13. What are the anticipated needs in 2 years, 5 years, 10 years, and 20 years?
14. What are the maintenance requirements?
15. Are there any sanitation problems?
16. Are the facilities accessible to the special-needs students?

Based on the responses to the above questions, the program needs should come in better focus, and therefore help lead to decisions about the requirements for the facility design. This input can then be balanced with the curriculum itself.

If a preliminary decision needs to be made whether to keep the program a) in the present facility, b) to remodel, or c) to construct new facilities, one question may be "Will the current facility hinder the quality of instruction?". An assessment can then be made to help determine some of the restraints of the existing facility. As this information is gathered it is appropriate to call upon outside consultants such as administrators, other teachers, parents, students, architects, and contractors to help in determining the feasibility of renovation or reconstruction. These individuals should provide input to help determine if the old structure is in sound structural condition and the cost effectiveness of renovation. On the other hand, they may locate so many structural barriers that it may become too costly to renovate.

Another factor this study may bring to light is the educational effectiveness of the existing facility. It is possible that structurally the existing area is satisfactory, but because of a variety of physical restraints, it is not educationally sound to attempt to use the facility for a manufacturing program.

Designing a new facility has the advantage of being able to use the latest technology and build a manufacturing laboratory that is as up-to-date as

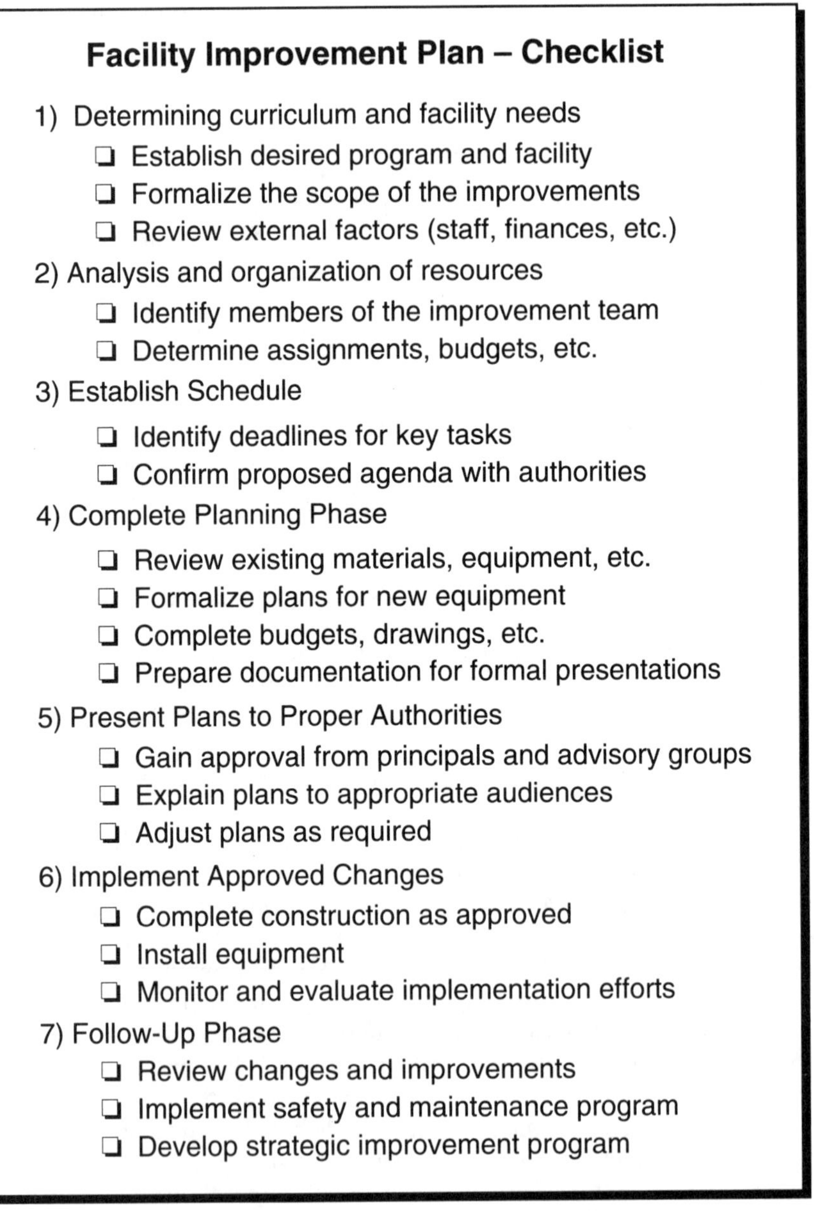

**Facility Improvement Plan – Checklist**

1) Determining curriculum and facility needs
   - ❑ Establish desired program and facility
   - ❑ Formalize the scope of the improvements
   - ❑ Review external factors (staff, finances, etc.)
2) Analysis and organization of resources
   - ❑ Identify members of the improvement team
   - ❑ Determine assignments, budgets, etc.
3) Establish Schedule
   - ❑ Identify deadlines for key tasks
   - ❑ Confirm proposed agenda with authorities
4) Complete Planning Phase
   - ❑ Review existing materials, equipment, etc.
   - ❑ Formalize plans for new equipment
   - ❑ Complete budgets, drawings, etc.
   - ❑ Prepare documentation for formal presentations
5) Present Plans to Proper Authorities
   - ❑ Gain approval from principals and advisory groups
   - ❑ Explain plans to appropriate audiences
   - ❑ Adjust plans as required
6) Implement Approved Changes
   - ❑ Complete construction as approved
   - ❑ Install equipment
   - ❑ Monitor and evaluate implementation efforts
7) Follow-Up Phase
   - ❑ Review changes and improvements
   - ❑ Implement safety and maintenance program
   - ❑ Develop strategic improvement program

*Figure 8-3: A sample facility improvement plan checklist.*

possible considering the constraints of finances, time, and personnel. During times of tight school budgets, it may be difficult to convince the taxpayers that a new facility is cost effective. Whether the decision is to build a new facility or extensively remodel the current facility, the major focus should be on students and supporting an approved curriculum.

## Analyzing Resources

Planning a manufacturing facility requires the efforts and coordination of a variety of resources. If the resources are not available or if certain individuals fail to accomplish the identified tasks, the end product may not totally meet the original objectives established for the program. Typical resources include people, equipment, and financial support.

*People.* All the individuals involved in the creation of a manufacturing facility, including the educators and the various design professionals, constitute the design team. This team should be aware of the philosophy and curriculum documentation. Then, they can cooperatively arrive at consensus on the most favorable facility plan that remains within the constraints of the available resources. The design team often consists of two subgroups: the educational team and the facility design team. The educational team is responsible for the establishment of goals and objectives for the program and determines how these should be carried out in the facility. This team should consist of administrators, teachers, educational consultants, and program advisory members. The facility design team is responsible for creating the physical facilities best suited to meet the established program needs. The facility design team would consist of architects, engineers, designers, contractors, and special consultants. After any preliminary plans are approved, various contractors will become a major part of the overall planning team.

*Equipment.* Equipment for a modern manufacturing facility will vary by the educational level of the students using the facility. In general, the equipment required at the elementary level will primarily consist of hand tools that match the age and abilities of the student. The equipment should be quality tools and not simply toys. The students need to realize at a young age that quality tools and equipment are needed to do a job correctly. Further, the care and proper use of tools is part of the total educational program.

The middle-school program requires hand tools, power tools, and larger equipment. The equipment should also be of high quality. Power equipment may be of the bench-top variety. Floor-mounted equipment reduces the flexibility of the manufacturing program and is generally not required because large items are generally not manufactured. At the senior high school and university levels there is a need to have standard-size equipment for a large variety of processing areas and fields. Extensive communication equipment, both hardware and software, are needed to allow the students to design and operate a full-blown manufacturing program from the initial idea stage to the production of the products to the dissolving of a student enterprise.

*Money.* Unfortunately, finances often play the most important role in the development of educational facilities. It is vital that the importance of the program be well established before any attempts to solicit funding. The program philosophy and manufacturing curriculum must directly parallel the goals of the institution. The funding issue can best be addressed only after these connections are in place. Every effort should be made to get the greatest value for the expenditures but not compromise on facility considerations that impact individual safety, educational goals, and the proper operation of the completed facility.

If an extensive program revision is accompanying the facility renovation program, financial resources for facilities and equipment may be further diluted. The entire funding package may include budgetary items such as staff in-service, consultant fees, curriculum development, maintenance services, etc. The actual budget for facilities and equipment may be a fraction of the total effort.

If the program is well conceived and documented, justification for the ideal facility (as originally designed) will be easier. The school and community will support a quality program if they understand the positive impact the program will have on their youth.

## Establishing a Schedule

Facility improvement plans take time to design, gain approval, implement, and bring on-line. A clearly defined schedule should be established early in the planning process. This phase should result in deadlines for each task in the development efforts.

Major facility conversion or construction projects must often be conducted during the summer months when classes are not in session. Therefore, the planning efforts of the design team must match the time-line of the school itself. Curriculum planning, needs assessments, formal presentations, budgetary reviews, and similar efforts should be completed within the established schedule. Variations from the initial time frame may jeopardize an entire improvement project.

A special mention should be made of the factor of *time* in the facility design process. Although not listed as a resource in the previous section, time is often overlooked as an essential component in larger projects. As the planning, design, and implementation efforts continue, time may become perhaps the most precious of all resources.

## Presenting Developed Plans for Approval

Facility enhancement projects usually result in a formal presentation to higher authorities to seek final approval of the developed plans. This process also applies in designing educational facilities. Gaining approval for a renovated or newly constructed manufacturing area will likely involve several steps.

The project starts at the departmental level, is reviewed by the administration, and typically ends up at a school board meeting for final action. Along the way, advisory committees and building principals may offer constructive criticism. Outside consultants may voice relevant concerns and suggestions.

All presentations should be completed in a professional manner. The dress of all presenters plus handouts, graphics, and supplemental materials should be appropriate for the audience. The presenters should practice their talk to polish the presentation and insure the address will fit the allowable time. Questions should be fielded in an honest and thoughtful manner.

Following each stage in the review process, modifications may be required of the entire plan or specific parts of the proposed project. Make all suggested corrections and improvements immediately and continue through the approval process.

## Implementation and Follow-Up

The goal of facility improvement efforts is to develop a more effective instructional facility to support the manufacturing sequence. Once the facility plans have been approved, the plans are implemented. This marks a period of construction with the accompanying dust and dirt, heavy traffic, the movement of large materials and equipment, and supplies littering the scene.

One individual (e.g., a manufacturing teacher, principal, or head of building maintenance) should monitor all construction efforts. Specified plans may be altered slightly during the renovation work so someone on-site should be available to respond to minor changes. Examples of critical decisions that may be needed on short notice include the following:

- Determining temporary locations of larger equipment as they arrive.
- Handling paperwork of all supplies, materials, and equipment as they are delivered to the site.

- Protecting new computer equipment from dust, water, etc.
- Moving existing equipment and materials that are to remain as inventory in the new facility.
- Point out problem areas to construction workers still on the site.

Follow-up efforts are important to conclude the original project and establish future development of the technology program. Once the new facilities are complete, plan to invite school officials to the department to "showcase" the improvements. Numerous details must be taken care of including a) the returning of warranty forms for new equipment, b) formalizing safety plans for the revised facility, c) updating maintenance programs, and d) related tasks.

# SPECIAL CONSIDERATIONS OF THE MANUFACTURING ENVIRONMENT

A manufacturing program must exist within a total school environment. The uniqueness of the manufacturing sequence and its associated laboratories provide students with an opportunity to work with many of the tools, machines, materials, and processes of our technological society. Just as the coursework will vary at each level of the school system, so must be facilities to support the instruction. This section will address the varying requirements of the manufacturing environment.

## The Elementary Program

To allow the K-5 student to gain a better understanding of our technological society, it is important to provide the tools, materials, and environment for the child to explore technology. An interdisciplinary approach to the study of manufacturing and technology in the elementary school is suggested. Manufacturing activities should be introduced at this level to enhance the student's technological literacy and understanding. Students will best learn about manufacturing when its study is integrated with other subjects and used to reinforce other content.

At this level, children become aware of the large variety of manufactured items that we rely on every day. They should be able to identify the resources, the major processing systems, and their impacts on society and the environment. The equipment for manufacturing activities does not have to be sophisticated or expensive. Rather, it should be selected so it can be safely and conveniently used by younger students who have small hands, limited coordination, and minimal strength. The materials should be those that are easily processed with simple tools. Eye/hand coordination is not highly developed at this age, so close tolerances are not possible. Careful attention will be required to schedule, maintain, and use the tools and materials.

Additional facility suggestions for the elementary manufacturing program are outlined in Chapter 3. The chapter includes a brief discussion related to portable tool carts, work surfaces, and equipment. A supply and tool list is also noted.

## The Secondary Program

Manufacturing experiences at the middle school level should provide for the gradual transition from the typical self-contained elementary classroom to the highly departmentalized high school. Furthermore, the middle school curriculum should provide an environment where the student, not the program, is the most important and where the opportunity to succeed is fostered. Manufacturing activities taught at the middle school level typically address this criteria. In contrast, the high school program involves activities that place an increasing amount of responsibility on the learner.

The middle school facility should allow the students an opportunity to explore a variety of manufacturing processes. Depending on the size of the school, the program may require one or several full-time teachers. The middle-school philosophy of an exploratory, interdisciplinary, and vertically-integrated curriculum fits the learning style of these young learners.

The general middle school laboratory might include a wide diversity of equipment and materials grouped around major organizers of technology (i.e., construction, manufacturing, etc.). The facility should be flexible in nature to promote the integration of other disciplines plus enhance the student's overall understanding of their technological world. A significant amount of problem solving should be employed in activities conducted within the manufacturing section of the laboratory. The equipment needs to be of good quality, but it is not necessary to have larger equipment of an industrial scale.

Middle school and high school laboratories should not be structured around the use of a single material. Multiple materials should be available so students will be able to solve their individual manufacturing problems by selecting from a large variety of materials having different properties.

The high school program should provide an environment for the student to take on more self-directed studies as well as become a productive member of a group. The students should also develop a better understanding of their own abilities, interests, and aptitudes as they begin to choose a career.

In contemporary technology education programs, senior high school facilities should be structured to support approved curriculum and the courses selected by the district. The manufacturing laboratory may be a separate laboratory, part of a production laboratory, or part of the entire technology education facility. Ideally, it would be a separate facility totally dedicated to the manufacturing cluster. Tools, equipment, materials, and processes should be broad in scope and yet provide an opportunity for students to conduct independent studies in some depth in areas of interests.

The equipment available at the secondary level will need to be current, safely maintained, properly sized for the setting, and of good quality. The products manufactured will vary from year to year, and consequently, this variation presents a challenge when specifying an equipment list that would fit all programs and all levels. On the other hand, general criteria such as flexibility, mobility, etc. can be used to support the selection of equipment for the laboratory areas. To be able to process as many materials as practical, equipment should be selected for its capability to handle standard separating, combining, forming and conditioning tasks. Since the manufacturing program is based on teaching the concepts of manufacturing and not so much on the production of sophisticated products, tabletop units are often more desirable than larger industrial-scale equipment. Finally, it is important to select equipment for the planned activities, not the individual interests and hobbies of the school personnel.

The basic equipment and special devices needed at this level include:

- Basic stationary power equipment to process wood, metal, plastics, etc.,
- Standard set of hand power tools,
- Computers and printers,
- CAD, CAM, and simulation software,
- Simple robotic and CNC equipment,
- Basic welding equipment,
- Foundry area and equipment,
- Material testing apparatus,
- Packaging equipment,
- Equipment to prepare simple graphic media,
- Photocopier,
- Quick-action clamps,
- Material handling equipment, and
- Tabletop material processing equipment.

All facilities should be set up to include clean, quiet areas as well as areas where considerable amount of smoke, chips and sawdust are generated. Naturally, computers and related equipment need to be part of the total manufacturing experiences. This equipment is best kept away from dirt, dust, and similar substances. Computers and automated equipment are often placed on carts for portability plus the ability to move the devices away from harmful substances.

## Teacher Education Programs

Teacher education facilities should reflect the latest in manufacturing technologies. A variety of laboratories clustered together enable the students to integrate the clusters and the professional teacher preparation sequence. In addition to the manufacturing lab, there should be a variety of support areas such as seminar rooms, resource centers, computer labs, etc. Again, the facility needs to be flexible to allow for maximum utilization of space and future growth.

The collegiate program has a dual purpose: a) to provide education to college or university students so they can better understand manufacturing systems and b) to provide an educational background and experiences for future manufacturing teachers. The manufacturing area within the curriculum is important and requires a facility designed to provide the future teachers with the necessary skills to be laboratory competent. Ideally, the total technology education facility should be in a separate facility equipped with up-to-date equipment and tools that can be used to bring about technological literacy in the students enrolled in the college or university. However, equipment requirements at the technology teacher education level will vary depending upon the curriculum structure of the collegiate program.

## Manufacturing Facilities and Public Relations

Naturally, a manufacturing facility is important for all supporting program goals and activities. One "hidden" agenda might involve public relations. The importance of the facility comes into play when individuals are interacting with the facility. This includes students, other teachers, guidance personnel, administrators, parents, and members of the community.

The transition to technology education has resulted in much confusion among those in the educational community. Many individuals fail to understand the nature of modern technology-based programs. They may be better informed after a visit to a well-designed and maintained technology education facility.

Facility design plays an important role in public relations and promotional activities. The visual appearance of a department and activities taking place in the area can either create interest and curiosity or immediately close the visitor's mind. It is imperative the manufacturing facility be an attractive, functional area that promotes the program in a positive manner.

The overall appearance of the facility goes a long way in promoting the manufacturing program. Place yourself in the role of a student and ask the simple question "Would I like to work and study in this environment?". The appearance of the physical setting establishes a lasting impression that can assure students and visitors leave the facility with a good impression of the manufacturing program.

## SUMMARY

Manufacturing facilities provide the educational environment for learning about the dynamics of modern manufacturing. Depending upon the nature of the program, manufacturing activities may be conducted within the regular classroom or in dedicated laboratory areas. This chapter outlined suggested facility requirements for programs at the elementary, middle school, high school, and collegiate levels.

An established curriculum is essential before proposing a facility improvement project. The new facility should support the approved curriculum. A model for improving facilities was presented to assist in renovating existing laboratories or designing new areas. The major steps included analyzing the available resources, setting a schedule and scope, completing all planning efforts, gaining approval of the proposed changes, implementing the modifications, and follow-up plans.

# *REFERENCES*

Braybrook, S. (Ed.). (1986). *Design for research: Principles of laboratory architecture*. New York: Wiley Interscience.

Broadwell, M. M. (1979). Classroom Instruction. In R. L. Craig (Ed.), *Training and development handbook: A guide to human resource development* (pp. 33-1 through 33-13). New York: McGraw-Hill.

Daiber, R. A., & LaClair, T. D. (1986). High school technology education. In R. E. Jones & J. R. Wright (Eds.), *Implementing technology education* (pp. 95-137). Encino, CA: Glencoe Publishers.

Erekson, T. L. (1981). *Accessibility to laboratories and equipment for the physically handicapped: A handbook for vocational education personnel*. Springfield, IL: Illinois State Board of Education, Division of Vocational and Technical Education.

Gemmill, P. R. (1979). Industrial arts laboratory facilities: Past, present, & future. In G. E. Martin (Ed.), *Industrial Arts Education Retrospect, Prospect*, Bloomington, IL: McKnight.

Gould, B. P. (1986). Facilities programming. In S. Braybrook (Ed.), *Design for research; Principles of laboratory architecture* (pp. 18-48). New York: Wiley Interscience.

Hales, J. A., & Snyder, J. F. (1981). *Jackson's Mill industrial arts curriculum theory*. Charleston, WV: West Virginia Board of Education.

Helsel, L. D., & Jones, R. E. (1986). Undergraduate technology education: The technical sequence. In R. E. Jones & J. R. Wright (Eds.), *Implementing technology education* (pp. 171-200). Encino, CA: Glencoe Publishers.

Henak, R. (Ed.). (1989). *Elements and structure for a model undergraduate technology teacher education program*. Council on Technology Teacher Education, Reston, VA: International Technology Education Association.

Huss, W. E. (1959). Principles of laboratory planning. In R. K. Nair (Ed.), *Planning industrial arts facilities* (pp. 48-65). Bloomington, IL: McKnight.

James, R. W., & Alcorn, P. A. (1991). *A guide to facilities planning*. Englewood, NJ: Prentice Hall.

Johnson, F. M. (1979). The organization and utilization of a planning committee. In D. G. Fox (Ed.), *Design of biomedical research facilities; Proceedings of a cancer research safety symposium*, Report No. 81-2305 (pp. 3-16). Bethesda, MD: National Institutes of Health.

Kemp, W. H., & Schwaller, A. E. (1988). Introduction to instructional strategies. In W. H. Kemp & A. E. Schwaller (Eds.), *Instructional strategies for technology education* (pp. 16-34). 37th Yearbook of the Council on Technology Teacher Education. Mission Hills, CA: Glencoe Publishers.

Lauda, D. P., & McCrory, D. L. (1986). A rationale for technology education. In R. E. Jones & J. R. Wright (Eds.), *Implementing technology education* (pp. 15-46). Encino, CA: Glencoe Publishers.

Lewis, H. F. (1951). Site selection, design, and construction. In H. S. Coleman (Ed.), *Laboratory design* (pp. 75-79). New York: Reinhold.

McHaney, L. J., & Berhardt, J. (1988, September). Technology programs: The Woodlands, Texas. *The Technology Teacher*, *48*(1), 11-16.

Polette, D. L. (Ed.). (1991). *Planning technology teacher education learning environments*. Council on Technology Teacher Education. Reston, VA: International Technology Education Association.

Rokke, D. L. (1988). *An overview of technology laboratory design*. Paper presented at the annual conference of the American Vocational Association, St. Louis, MO.

Rokusek, H. J., & Israel, E. N. (1988). Twenty-five years of change (1963-1988) and its effect on industrial teacher education administrators. In D. L. Householder (Ed.), *Industrial teacher education in transition: Proceedings of the 75th Mississippi Valley industrial teacher education conference* (pp. 219-244). College Station, TX: Mississippi Valley Industrial Teacher Education Conference.

Seal, M. R. & Goltz, H. A. (1975). Planning principles. In D. E. Moon (Ed.), *A guide to the planning of industrial arts facilities* (pp. 55-64). Bloomington, IL: McKnight.

Shackelford, R., & Henak. R. (1982). *Making industrial education facilities accessible to the physically disabled*. Reston, VA: International Technology Education Association.

CHAPTER 9

# Synthesis Of Systems/Approaches For The Study Of Manufacturing

Dr. Richard D. Seymour (Associate Professor)
*Ball State University, Muncie, IN*

Feather (1989) reminds us that the "Industrial Age spawned the still prevalent mass-production model of education—factories of learning" (p. 100). Of course, manufacturers can no longer compete using the techniques and management styles of the 1800s. In much the same way, technology educators must alter their program goals, actions, and course offerings to follow the agenda of the modern society. This means improving instructional strategies, individual courses, professional efforts, and facilities to support new goals.

The programs outlined in this yearbook reflect decades of curriculum development in the area of manufacturing technology. The primary momentum of modern manufacturing coursework evolved from curriculum projects such as the American Industry Project, the Maryland Plan, and the Industrial Arts Curriculum Project (IACP). Lux (1981) estimated that the IACP "World of Manufacturing" was fully introduced into 15 percent of the schools in the U.S. during the 1960s, with parts of the curriculum adopted into numerous other industrial arts programs. The Stanley Tools Mass Production Contest promoted the study of production technology and enterprise topics during the early 1980s. In addition, a significant number of state and local initiatives resulted in manufacturing topics being included in public schools, Figure 9-1.

Today, manufacturers as well as manufacturing education stand at an important crossroads. Both institutions face similar issues—poor products/outcomes, lack of foresight, inefficiency, financial problems, erratic support from government, and a serious lack of internal leadership. Both spend considerable energies defending their practices and outcomes. Both need to

**Significant Developments in Manufacturing Technology Education**

1960s – IACP World of Manufacturing
American Industry Project

1970s – Stanley Tools Mass Production Contest
*Manufacturing Forum* magazine

Early 1980s – Jackson's Mill Curriculum Project
Technology Education Symposium Series
Industry & Technology Project
State Curriculum Projects (New York, Illinois, Indiana)

Late 1980s – TECA/SME "Live" Manufacturing Contest
Center for Implementing Technology Education (CITE)
Numerous State and Local Projects
SME/CITE Manufacturing Guides

1990s – Elementary School Manufacturing Programs
Math/Science/Technology Projects

*Figure 9-1: Major developments in the evolution of manufacturing programs in technology education.*

place more attention on developing solutions to existing problems and less time on documenting the reasons for current inadequacies.

Many social changes have combined to alter the workplace. For instance, the desire to have more leisure time combined with a general reduction in work force needs have led to flexible scheduling and extended vacation hours. The increasing pace of product introductions has created a demand for adaptable, educated workers willing to help in multiple positions. Gerelle & Stark (1988) point out that "women and minority groups, traditionally constrained to work at the lower echelons of companies, are demanding equal opportunity at a level that fairly reflects their skill, education, and experience" (p. 11). Additionally, management procedures have been changed to match new personalities and work ethics among labor, managers, and top executives.

Modern technologies have also altered the nature of manufacturing tasks. While automated systems, computer controls, and vision technology have contributed to efficiency and productivity, they also change the production environment. Cohen and Zysman (1987) define the prominent differences in the workplace when they note:

> Production work has changed. People go home cleaner; more and more of them leave offices rather than assembly lines. Service activities have proliferated. The sociology of work and the organization of society have changed along with the technologies of product and production. (p. 26)

Social changes have also accentuated the recent problems in the educational system. Numerous reports cite a drop in SAT scores, high dropout rates, low teacher pay, and poor reading and writing skills among graduates. Also, schools are suffering through times of budgetary cutbacks, larger class sizes, and continually higher expectations. Members of the industrial sector have routinely criticized teachers, administrators, and school boards for failing to adequately prepare young students for their roles as citizens and workers.

From a societal perspective, young students are often confused on educational strategies. The MIT Commission on Industrial Productivity (Dertouzos, et al, 1989) pointed out how "the individual does not know where [he/she] will be employed and does not want to invest time and money in acquiring skills that will become worthless" (p. 21). Since most individuals will hold numerous jobs throughout their productive careers, they often appear undecided concerning what academic programs might best prepare them for their future.

This chapter will discuss the current trends and issues in manufacturing programs at the public school and collegiate levels with a look toward the future. The first section covers the value of manufacturing technology education while the following section will review options for technology teachers. The final segment will deal with existing issues for enhancing the manufacturing sequence.

# THE VALUE OF MANUFACTURING EDUCATION

As discussed throughout this yearbook, manufacturing is often viewed using a systems approach (Colelli, n.d.; Lauda & McCrory, 1986; Wright & Shackelford, 1990; Savage & Sterry, 1990). The number and classification of inputs to a manufacturing system may vary, but it is generally agreed that human resources represent the most important input. Dertouzos, et al (1989) simply state "that the ultimate resource of an industrial economy is its people" (p. 21). Cohen and Zysman (1987) expand the point to include the notion that "labor as a factor of production is not just people, but people with particular skills, attitudes, and habits" (p. 228).

When a major institution such as the goods-producing sector is in trouble, society begins to look for reasons, causes, and options. The attention frequently returns to the general education provided by the public schools. An explanation of the important relationship between education and manufacturing is provided by Grayson, Jr. & O'Dell (1988):

> Education is directly linked to competitiveness. No society can have high-quality outputs without high-quality inputs. Education provides "human capital" to combine with "physical capital" for increased productivity and quality. That has always been true, but it is even more true in a global technological society. (p. 270)

The stated goals and objectives for contemporary schools have become a comprehensive listing of social, political, and economic outcomes of ambitious proportions. The list may have become too diverse for anyone (even dedicated educators) to adequately address. Further, many of the goals are seen as too general, vague, or unobtainable. Goodlad (1984) addressed the "we want it all" attitude and offered a rather narrow list of goals for schools that feature a) academic goals, b) vocational goals, c) social, civic, and cultural goals, and d) personal goals. Goodlad defends his list of five basic academic skills with the following observation: "In our technological civilization, an individual's ability to participate in the activities of society depends on mastery of these [basic] skills and [fundamental] processes and ability to utilize them in the varied functions of life" (p. 51). Interestingly, many of Goodlad's academic and social recommendations are addressed in the proposed coursework outlined in chapters 3-5.

Many new curriculum models have been directed by the three R's, variations of the math/science/technology theme, and a return "to the basics." But it is imperative that all curriculum development efforts should be created with a continuous eye on desired outcomes. Educators have many more challenges than just insuring that students learn to read, write, and master simple mathematics. Preparation for life and work should include a variety of instructional experiences while students complete their K-12 education. The topic of manufacturing appears to enhance the elementary and secondary experience since it addresses most of the themes of modern education.

When structuring an effective manufacturing curriculum, one challenge involves developing the same look toward future outcomes as is used in other disciplines. Technology education has been positioned in schools as *general education*, or education vital for all students. The program structures discussed in the initial chapter and chapters 3-5 in this book support this premise. Content outlined at each level demands students be involved in activities that support their general knowledge of a technological society. Further, the focus is continually on the important theme of understanding technological devices, systems, and impacts. Feather suggests four basic proponents of education in the future, one of which is "technological skills, especially the ability to understand the capability of modern technology" (p. 99).

Manufacturing experiences, like all general education programs, should be structured to help all citizens complete their knowledge of our technological society. Grayson, Jr. & O'Dell suggest that schools focus on developing students with both good cognitive skills (thinking, content, etc.) and affective skills (interpersonal skills, flexibility, determination, etc.) since "an effective citizen or employee needs both" (p. 272). The goal is to increase the general abilities of all students, thus preparing them for multiple positions as consumers, voters, and productive workers.

Many educators and futurists agree that "the basics" of the productive citizen beyond the year 2000 will not be the same as the basics of previous decades. But, Naisbitt and Aburdene (1985) suggest that the individual and group skills related to leadership, creative problem solving, entrepreneurship, rational thinking, and communication appear to be sound for the coming generations. The action-based learning in modern manufacturing courses promote these important themes in a fun, dynamic environment.

## Educational Reactions to Global Changes

International strength and influence is typically dependent on a country's industrial base, and the fundamental component of industry involves people. As Cohen and Zysman put it, "unless we have the [human] skills to employ the new production possibilities, no amount of investment capital will make us competitive with countries that have invested in human as well as physical capital" (p. 229). Since people are more important than other inputs (raw materials, energy, money, etc.), educators have been challenged to better prepare our future citizens and workers.

Educators will likely face numerous suggestions over the coming decade to overhaul the institution. Proposals such as raising graduation requirements, lengthening the school year, implementing accreditation procedures and teacher certification efforts, etc. will dominate the headlines. Other major issues may involve a combination of liberal arts and vocational ventures. In developing new policies and strategies, schools must review the realities of current and future learners and adapt to their aspirations and needs. Toffler (1990) reminds us that emerging changes "go beyond questions of budgets, class size, teacher pay, and the traditional conflicts of over curriculum" (p. 368).

New actions to improve education should be thoroughly reviewed by school, business, and civil leaders. One example is the popular call of lengthening the school year to bolster academic performance. Numerous studies have called for longer school days and years to better academic performance. But, a local manufacturer might suggest that if a specific system doesn't work, the best solution is to fix the existing system rather than increase capacity or adjust the current schedule to produce more quality products. Likewise, requiring students to increase their participation in an already "poor" educational experience may not be the best solution.

Feather (1989) explains recent attempts to "futurize the curriculum" in suggesting new teaching methods be used to prepare students to solve complex problems, increase math/science/technology understanding, enhance decision making skills, become better able to work cooperatively, and (perhaps most importantly) learn how to learn. He also observed:

> Above all, we must place an emphasis on "learning to learn," whereby people can find their way through ever more complex labyrinths of information. There is too much to learn. We need, as a basic principle, to know where to find knowledge . . . . and learn to manage and apply knowledge in long-term planning and decision-making. (p. 97)

Examples of improving a student's ability of "learning to learn" are found throughout the manufacturing cluster. For instance, research and development activities address an individual's ability to fully understand a problem or opportunity, then develop creative, practical solutions based on new and applied information. Laboratory assignments often challenge students to learn new information during both the initial research phase and throughout the development tasks. Problem-solving activities also promote this concept of "learning on your own" when specific problems are introduced and later when students propose creative solutions. Again, learning to learn is "considered the most basic of all skills" (Boyett & Cohen, p. 279).

Manufacturing educators have placed more emphasis on open-ended problem solving that requires creativity and divergent thinking. Waetjen (1989) notes how modern "curriculum guides identify problem solving as a major teaching technique for enhancing students' learning" (p. 1). Merli (1990) also suggests that topics such as problem identification, problem solving, etc. become an essential part of future learning. Examples throughout chapters 3-5 outline possible problem-solving themes during design, development, and R & D activities.

Virtually everyone points to the demand for a cooperative attitude among workers and citizens in a technological society. Lately, schools have integrated disciplines and experiences to enhance the education of young learners. Teachers often convert their individual assignments into small group activities that require knowledge from several disciplines to solve the problems. Again, this trend is represented in many of the activities outlined in the previous chapters of this yearbook. For instance, the major theme of a manufacturing enterprise activity is to stress the importance of collectively working as a unit to address daily problems and opportunities.

Environmentalism has become an important agenda in most nations and many industries. The "green wave" often appears in educational programs when the consequences of social and technological actions are assessed. Teachers typically approach the topic of recycling when discussing packaging and consumer goods. Some manufacturing programs are beginning to address the concepts of scrap versus waste in selecting specific processes and systems.

Instruction in technology-based coursework at the public school level is currently being shaped by many forces. For instance, former industrial arts teachers have introduced a wide range of classes including production systems, design and ergonomics, and bio-technologies. At the same time, new calls for Tech(nology) Prep(aration) and workforce literacy have come from legislators and industrialists. School boards have long sought to define and implement exclusively vocational courses at the higher levels (comprehensive high schools, "2+2" programs, etc.). In addition, area industries often impose self-serving expectations on community schools (Harmon, 1992).

## Business and Industrial Support

Both the industrial sector and educational community share common problems and can often use similar initiatives to promote improvement. But each institution is also different and one must realize the distinct variations. Schools deal with young individuals that have varying levels of motivation and understanding, and develop at different physical and intellectual rates. More importantly, young learners cannot be standardized to the same extent as raw materials, manufacturing processes, and consumer products. Therefore, programs that work in a particular industry might not be appropriate in helping area schools.

Still, business and industrial leaders must team up with educators to create better schools. Contributions of time, expertise, materials, etc. are welcomed by teachers and administrators. While money is also appreciated, other donations are sometimes more useful. It is important to note that resources considered insignificant to manufacturers and commercial outlets often greatly enhance a local school program.

Obviously, manufacturing teachers should encourage partnerships with area industries for instructional and program purposes. Local firms can offer assistance by authorizing field trips or providing guest speakers. The manufacturing program can also benefit by using examples from local industries (quality control manuals, samples of plant layout drawings, old prototypes, etc.). Certainly, instructional efforts are more interesting if students see and use familiar materials from nearby firms.

Unfortunately, industrial support of the educational community is typically vague or self-serving (Reich, 1992). For instance, as companies negotiate tax advantages to contain overhead, they often hurt local governments, area schools, and various services. The cost-cutting measures may impact the local community in unexpected ways. Manufacturers should realize the most important resource in the community is the local school system and be supportive of the educational efforts.

## FUTURE EMPHASIS FOR MANUFACTURING EDUCATORS

This is an exciting time for manufacturing teachers. For one of the few times in history, industry, government, and society in general have all recognized the vast importance of education. Suggestions for improving instructional efforts at the elementary, secondary, and collegiate levels are being proposed and supported from multiple sources. Major curriculum and facility changes are imminent. Interestingly, many of these recommendations are already documented in manufacturing textbooks, teachers' manuals, and curriculum guides.

As outlined in Figure 9-1, manufacturing technology education is not new to the academic scene. Neither are the proposed teaching strategies, instructional media, facilities and equipment, or processes described in earlier chapters. What is unique is the opportunity to implement this coursework in public schools and the collegiate level with the full endorsement of administrators, school boards, department heads, and deans.

Manufacturing technology has evolved as the most developed of the four clusters originally proposed by the Jackson's Mill Curriculum Project. The field of technology education includes a wealth of textbooks, media, and related materials to support the study of manufacturing. Today, the major ingredient in the implementation of new programs is a dedicated, professional teacher to nurture the program. In many circles, this type of person is said to be an individual who can "champion" a cause.

This section will focus on the manufacturing teacher and review key elements in the successful implementation of manufacturing technology.

Topics will include the development of teaching skills, models of new content, and professionalism.

## Teaching Manufacturing in an Information Age

Educators are actually *managers* of instruction. Teachers routinely manage the classroom and laboratory activities, experiences, and behaviors of young learners. Anderson (1985, p. 45) reminds us that "as managers, we must use all the methods at our command" including instructional media, tools, supplies, and related resources. Still, the characteristics of teacher—personality, adaptability, knowledge, motivation, leadership, and similar traits—represent the essential element in manufacturing programs.

The general education program available in public schools is ideally suited to young learners; it establishes a foundation for future learning. But, Gerelle and Stark (1988) note that learning must always continue since "a general education is no longer a guarantee of employment" (p. 11). Just as software and equipment often become obsolete, human knowledge and abilities can also be out-of-date. Therefore, a manufacturing program should be dynamic with new content continuously introduced into the program.

Just as students are expected to "learn how to learn," manufacturing teachers must also be willing to continually upgrade knowledge and skills. In an age of continuous technological change, it is probable a manufacturing course will never be taught the same way twice. New content and instructional strategies might be incorporated into the curriculum each grading period or semester. Manufacturing teachers should know a) where key information is located, b) how to access it, and c) use it to achieve program goals. This may involve using enhanced library skills, using database software, receiving information via a modem, or being able to use CD-ROM systems.

Recently, the computer has become the most important piece of equipment in the manufacturing area. Besides its accepted use for word processing and machine control, modern applications include factory simulations, preparation of complex graphics (package design, plant layout drawings, tooling design, etc.), and production documentation. For instance, standard business forms can be loaded as a desktop accessory to help prepare inspection forms, process charts, and time study forms. Peripherals such as a bar-code reader illustrate the advantage of maintaining inventory with this technology. Students and teachers find that computers make many routine tasks more efficient and fun.

At all levels, it is essential the manufacturing teacher have a computer available for both educational and professional activities. Student handouts—tests, design briefs, laboratory recording sheets, daily forms, etc.—should be done on a computer and printed using a laser printer. Since many association activities involve frequent communication, computers and printers are helpful to promote professionalism. Finally, correspondence with parents and colleagues should be word processed and printed on a laserwriter to present a favorable image of the manufacturing program.

## New Curriculum Models

Recent curriculum models have forced the profession to think of key topics not addressed in present coursework. Perhaps the best known "new" model is from the Conceptual Framework project directed by Savage and Sterry (1990). This project formally identified the bio-related technologies (including agriculture and medical technology) as a major content area. The authors note how bio-technology "applies biological organisms to make or modify products" (p. 17). Two familiar examples of industrial processes include fermentation and distillation. Advancements in this area continue to emerge as a) new technologies are perfected and b) new uses are discovered for the new technologies.

Content models in chapters 4 and 5 of this yearbook discussed raw materials with an emphasis on solids. One suggestion to enhance the study of manufacturing at all levels is to incorporate activities related to liquids and gases (along with their respective processing techniques). For example, a unit on refinery operations would allow students to review the petroleum cracking process as well as explore obvious interdisciplinary implications with the math and science program. In another case, new bio-technologies could be covered during a unit on harvesting (e.g., sawing or cutting) or propagating (e.g., fermentation).

Perhaps the most significant recommendation from the Savage and Sterry project is that technology-based programs include a more global view of materials, processes, and systems. While the manufacturing sequence already includes many forms of resources, it would be appropriate to expand the focus and content of certain courses. Emerging practices, such as bio-technology, represent a new theme to add to the study of manufacturing technology.

While many curriculum models are content based, others have focused more on the *processes* of technological actions, Figure 9-2. Two major processes that have become major themes (and even course titles) involve design and problem solving. For instance, courses related to design topics are popular in certain schools. The British system of education is highly structured toward both design and problem-solving activities. All content and instructional activities follow specific models and students routinely

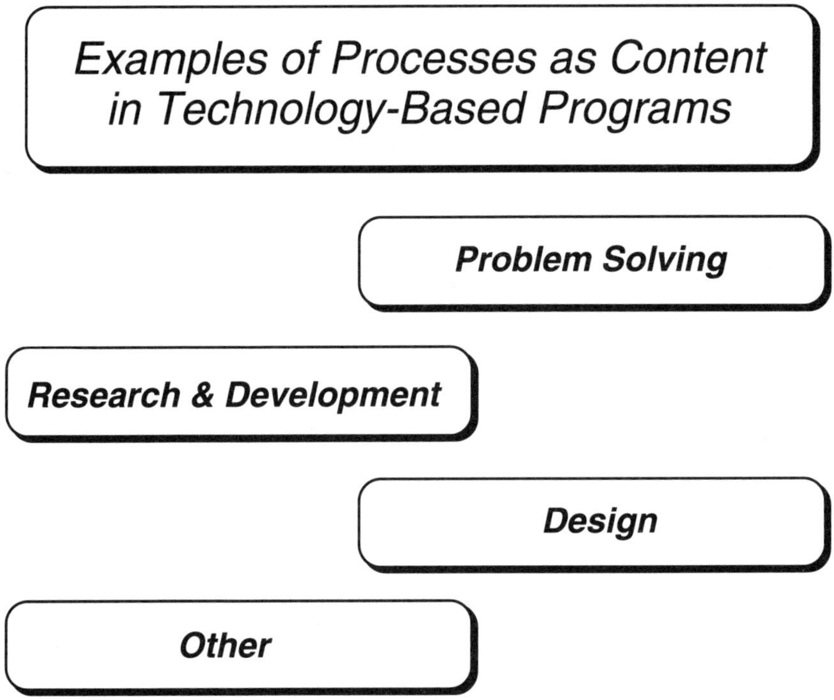

## Examples of Processes as Content in Technology-Based Programs

**Problem Solving**

**Research & Development**

**Design**

**Other**

*Figure 9-2: Many technological processes have become actual course titles. This chart includes several of the processes related to manufacturing that have major themes in modern technology education prrograms.*

trace their progress through the processes. Other programs feature curriculum models that address processes such as research, industrial engineering, and ergonomics.

Manufacturing teachers should select appropriate models for key processes in each course and provide instruction related to the tasks. As students complete their activities following proper techniques, each step should be documented to show progress and understanding. This helps insure the students realize they are following a structured technological process (problem solving, design, etc.).

## A Focus on Students

During the past few years, students in many secondary and collegiate programs have participated in manufacturing contests sponsored by local, regional, and international associations (Betts & Van Dyke, 1989). This popularity of student organizational activities signals a new time in technology-based programs. These activities reflect how students have become the solid focus of teacher's instructional and professional efforts.

Manufacturing teachers have a unique opportunity to shape the lives of all students in their programs. Unfortunately, it is often difficult to pay attention to individual students within the constraints of class activities and schedules. Besides, teachers and learners act differently in a formal classroom situation compared to interaction during extracurricular activities. Student clubs and organizational activities allow the teacher to work one-on-one with individuals in a fun, relaxed atmosphere. In addition, the competitive spirit associated with contests, tournaments, etc. can be motivational for students and teams.

The Technology Student Association (TSA) and Technology Education Collegiate Association (TECA) both sponsor manufacturing-based contests that allow students to participate in challenging, exciting competitions. For instance, the TECA/SME "Live" Manufacturing Contest involves collegiate teams designing and installing a full production system within a 4-hour period. Typically, 12-20 quality products are produced using the developed system with an emphasis on teamwork and productivity. Students are challenged to draw upon many talents throughout the work sessions and final production run. Similar contests for high school students are sponsored by TSA.

Student organizational activities represent an effective way to cover manufacturing topics outside the normal sequence of scheduled classes. For instance, production contests are popular with students, and they challenge learners to improve their understanding of various technologies. Students find the learning more meaningful when they can immediately apply their knowledge and skills in solving specific problems. Also, a cooperative spirit is fostered among team members.

But, the major advantage of organizational activities is the intellectual and social growth displayed by students. Since students are the focus on all instructional efforts, it is rewarding to see the energy and enthusiasm of young learners in situations beyond the normal classroom activities. Sponsorship of student organizations is a gratifying way for manufacturing educators to enhance student academic and social growth.

## PROGRAM ISSUES FOR MANUFACTURING EDUCATORS

Manufacturing technology is a dynamic area of modern society. Manufacturers are in a continuous fight to remain competitive in a global marketplace. In fact, the ability to master change has become an important trait of successful companies. Danforth (1985) reminds us that:

> To be successful now and in the years to come, a company must be able to handle change. . . . Though it ebbs and flows, change is a continuing process. It never stops. The strategic ability to adapt to economic, social and technological change determines whether a company thrives and grows or withers and dies. (p. 89)

The same statement is applicable to manufacturing education. Today, sufficient instructional materials, media, and curriculum guides are available to support manufacturing programs from the elementary through doctoral level. Perhaps the primary term is "today" since future technologies have not yet been invented, nor has the coursework to address future student needs been finalized.

Continuous change is upsetting to many educators, but is an important part of the context for studying technology. Many technological changes involve simple modifications in existing products, processes, or systems. Other changes are often referred to as "advancements" due to their obvious benefit to society. Advancement is considered to be synonymous with concepts such as progress, improvement, and contribution.

A program goal of manufacturing teachers should be to *advance* the entire program on a regular basis. Improvements in courses, activities, media, etc. are key to progress in education. Teachers need to make a commitment to maintain a quality program. This includes all aspects of the instructional system:

- Faculty skills and knowledge,
- Course offerings,
- Student handouts and forms,
- Instructional strategies,
- Delivery systems,
- Individual and group interaction skills,
- Rational thinking skills,
- Interdisciplinary activities,
- Equipment and supplies,
- Recruiting and public relations efforts,
- Instructor motivation and enthusiasm,

- Facilities, and
- Professionalism.

This is not an exhaustive list. Rather, it suggests the numerous details and efforts required to maintain a quality program. Young learners deserve nothing less.

Manufacturing teachers should plan on a lifetime of learning and exploration. For instance, adding a manufacturing enterprise activity to a technology program means new instructional strategies are appropriate. A workshop on cooperative group learning or on how to evaluate group efforts would be helpful. The addition of the enterprise experience might also result in a need for the teacher to seek further knowledge related to marketing systems, labor relations, or company finances. Again, the decision to teach manufacturing involves a commitment to individual professional (as well as program) improvement.

## Manufacturing Education as an Answer

Recent criticism of the educational sector has highlighted the importance of mathematics, science, and technology. Reports and studies continually cite low test scores and poor graduation rates when discussing the inadequacies of modern schools. While these indicators present a less than flattering picture of today's educational institution, they also fail to recommend solutions to future schooling.

Among the many reports critical of schools, it is disappointing that the role of a contemporary technology education is rarely discussed. Numerous government and industrial initiatives emphasize the phrase *technology* in the call for educational reforms. For instance, Feather (1989, p. 99) refers to "technological skills"; Harmon (1992, p. 103) fully describes "the education needed for our companies to become the superior manufacturers of the future"; and Boyett and Cohen (1991, p. 279) note how "the need for continuous learning already exists in the workplace of today." Yet, few publications and reports specifically point to a technology education program to help improve technological literacy.

This trend indicates a giant opportunity for manufacturing educators. While everyone attempts to reach consensus about current problems and their implications, manufacturing teachers should step forward with their developed curricula. Manufacturing education includes a refined series of instructional experiences as reflected by the courses and topics outlined throughout this yearbook. In contrast to the problems documented in education and industry, manufacturing courses have actual *solutions* to present and future concerns.

## SUMMARY

Many topics, courses, and instructional strategies have been presented in this yearbook to address the important topic of manufacturing technology. The focus of this last chapter included a summary of key approaches to introducing manufacturing content and activities into the elementary, secondary, and collegiate setting.

Several societal issues related to manufacturing education were discussed. One section addressed the relationship of modern industry and academic programs in manufacturing technology. Another segment of the chapter outlined the nature of teaching manufacturing in the information age. Also, recent curriculum models were reviewed.

The future of manufacturing education is in the hands of dedicated classroom teachers. Present instructional materials are useful in implementing a modern manufacturing program, but an eye should be kept on the future. Continuous improvement is essential to maintain momentum in this field.

# *REFERENCES*

Anderson, R. (1985). Managing diversification (Chapter 2). In J. M. Rosow (Ed.). *Views from the top*. New York: Facts On File Publications.

Betts, M. R., & Van Dyke, A. W. (Eds.) (1989). *Technology Student Organizations.* (38th Yearbook, Council on Technology Teacher Education). Mission Hills, CA: Glencoe Publishing.

Bloom, A. (1987). *The closing of the American mind*. New York: Simon & Schuster.

Boyett, J. H., & Conn, H. P. (1991). *Workplace 2000: The revolution reshaping American business*. New York: Dutton Books (Penguin Books USA).

Cohen, S. S., & Zysman, J. (1987). *Manufacturing matters: The myth of the post-industrial economy*. New York: Basic Books, Inc.

Colelli, L. A. (n.d.). *Technology education: A primer*. Reston, VA: International Technology Education Association.

Danforth, D. D. (1985). Handling change—The key to corporate survival (Chapter 5). In J. M. Rosow (Ed.). *Views from the top*. New York: Facts On File Publications.

Dertouzos, M. L., Lester, R. K., & Solow, R. M. (1989). *Made in America (The MIT commission of industrial productivity)*. New York: Harper Perennial.

Edmison, G. A. (Ed.) (1992). *Delivery systems: Teaching strategies for technology education.* Reston, VA: International Technology Education Association.

Feather, F. (1989). *G-Forces: The 35 global forces restructuring our future*. New York: William Morrow and Company.

Gerelle, E. G. R., & Stark, J. (1988). *Integrated manufacturing: Strategy, planning, and implementation*. New York: McGraw-Hill.

Goodlad, J. I. (1984). *A place called school*. New York: McGraw-Hill.

Grayson, C. J. Jr., & O'Dell, C. (1988). *American business: A two-minute warning*. New York: The Free Press.

Harmon, R. L. (1992). *Reinventing the factory II*. New York: Free Press (Macmillan, Inc).

Kupfer, A. (March 9, 1992). How American industry stacks up. *Fortune. 125* (5). pp. 33-46.

Lauda, D. P., & McCrory, D. L. (1986). A rationale for technology education. In R. E. Jones & J. R. Wright (Eds.) *Implementing Technology Education* (35th Yearbook, American Council on Industrial Arts Teacher Education). Encino, CA: Glencoe Publishing.

Lux, D. G. (1981). Industrial arts redirected. In R. Barella & T. Wright (Eds.). *An Interpretive history of industrial arts: The interrelationship of society, education, and industrial arts* (30th 1981 Yearbook, American Council on Industrial Arts Teacher Education). Bloomington, IL: McKnight.

Merli, G. (1990). *Total manufacturing management*. Cambridge, MA: Productivity Press.

Naisbitt, J. & Aburdene, P. (1985). *Re-inventing the corporation: Transforming your job and your company for the new information society*. New York: Warner Books.

National Center for Manufacturing Sciences. (1990). *Competing in world-class manufacturing: America's 21st century challenge*. Homewood, IL: Richard D. Irwin, Inc.

Reich, R. B. (January, 1992). Is big business support for education just hype? *Vocational Education Journal, 67* (1), pp. 62.

Rosow, J. M. (Ed.). (1985). *Views from the top*. New York: Facts On File Publications.

Savage, E. & Sterry, L. (1990). *A conceptual framework for technology education*. Reston, VA: International Technology Education Association.

Toffler, A. (1990). *Power shift*. New York: Bantam Books.

Waetjen, W. B. (1989). *Technological problem solving: A proposal*. Reston, VA: International Technology Education Association.

Weiss, J. M. (1987). *The future of manufacturing*. Red Bank, NJ: Bus Fac Publishing.

Wright, R. T. & Shackelford, R. L. (1990). *Product design and engineering*. Muncie, IN: Center for Implementing Technology Education.

Wright, R. T. (1990). *Manufacturing systems*. South Holland, IL: Goodheart-Willcox.

# *INDEX*

## *—A—*



## *—B—*



## *—C—*

## —*D*—

## —E —



## —F —

## —G—



## —H—

## —J—



## —K—



## —L—



## —M—

## —N—

## — S —



## — T —

## —U—



## —V—



## —W—